Water flow in plants

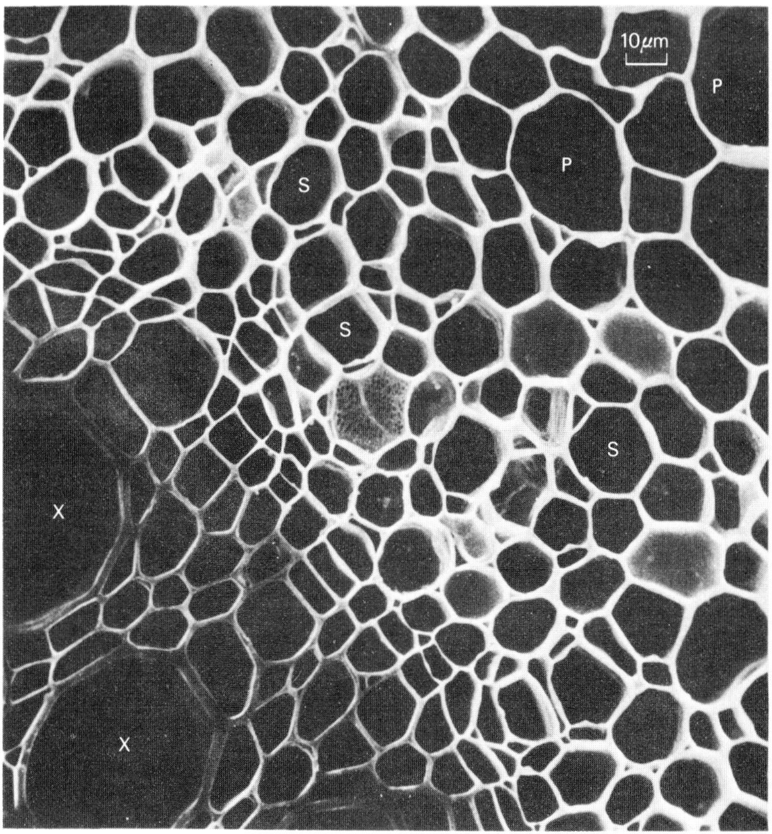

Frontispiece A thin handcut cross-section from part of a castor bean (*Ricinus*) leaf stalk viewed by scanning electron microscopy. Large xylem conduits with lignified walls (X, lower left) are embedded in xylem parenchyma cells. Centrally a sieve plate is shown surrounded by phloem parenchyma cells, severed sieve tubes (S) and the much smaller companion cells. Large cortical parenchyma cells (upper right) provide support. The difficulty of distinguishing between conducting tubes and supporting cells, with transverse walls removed by thin sectioning, is obvious.

Cover illustration Bordered pit from *Abies grandis* (Grand Fir) magnified about 20,000 times. The outer border has been fractured, leaving the pit membrane with torus behind. (Courtesy G. Puritsch and R. P. C. Johnson.)

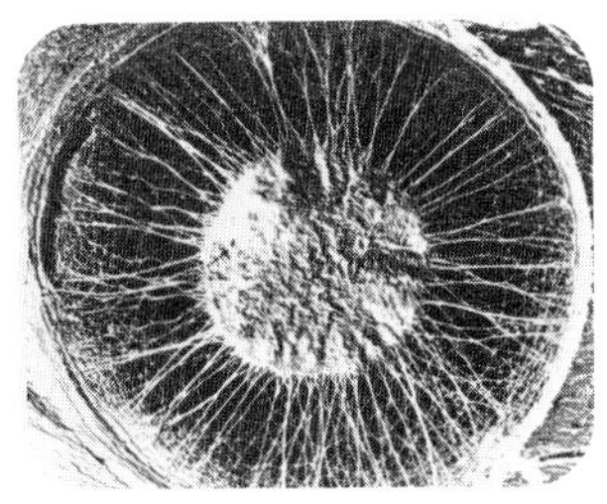

Water flow in plants

John A. Milburn

Department of Botany
University of Glasgow

Longman London and New York

Longman Group Limited London

Associated companies, branches and representatives throughout the world

Published in the United States of America by Longman Inc., New York

First published 1979

British Library Cataloguing in Publication Data
Milburn, John A
Water flow in plants. – (Integrated themes in biology).
1. Plant translocation
I. Title II. Series
581.1′13 QK871 77–30743

ISBN 0–582–44387–3

Printed in Great Britain by
Butler & Tanner Ltd, Frome and London

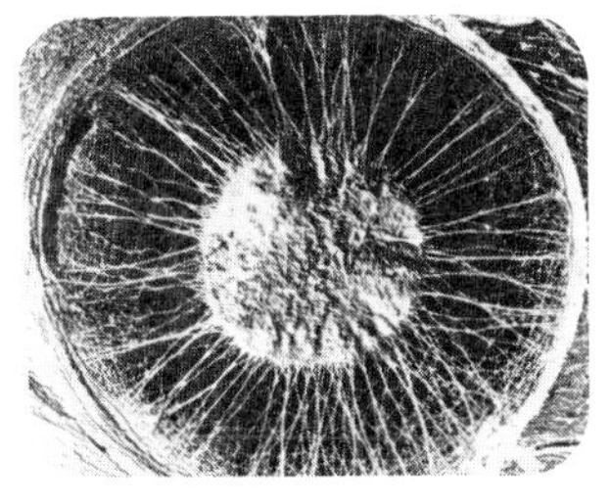

Contents

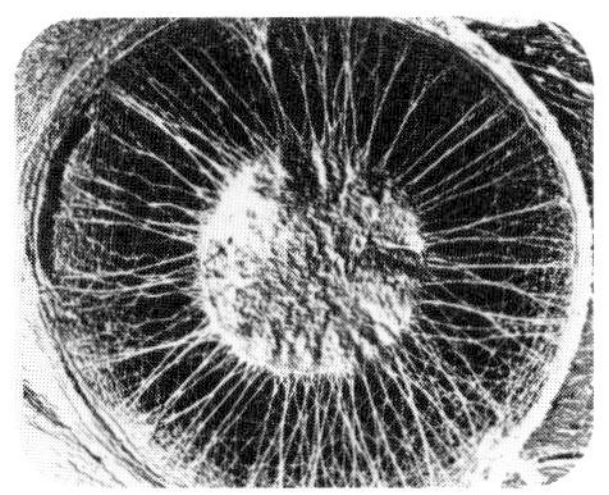

Preface

In this short text I have attempted to fulfil a strictly limited set of objectives to meet what seem to be the most general needs. These include an understanding of the basic mechanisms involved in water flow, current research progress, a means to grapple with quantitative problems and some of the applications in agriculture, forestry and horticulture. Much of the text is orthodox, but I have not hesitated to adopt more radical approaches where necessary.

Accordingly, the book begins, somewhat conventionally, with the properties of water, stressing the relative simplicity of the underlying principles. These are then illustrated by a series of interesting models and machines which provide an easy introduction to many physical processes in plants.

Since most of the mechanisms which have been elucidated by qualitative experiments are now quite familiar, plant physiology has become increasingly more quantitative in character. For this reason I have tried to include useful equations with examples illustrating their application in actual experiments. SI or SI-compatible units have been used throughout, because in my experience students find the mixture of units commonly used in current literature unnecessarily confusing. I hope students will be encouraged to regard quantitative equations as pieces of equipment into which their data can be fed to give answers to their queries. We still have a long way to go before we will be able to compare accurately, in quantitative terms, physiological processes in plants, but the trend is firmly established. From this quantitative approach we may expect many fascinating patterns and stratagems of plant behaviour to emerge in future.

Many new techniques have become available recently for quantitative studies. I have tried to explain the principles of these simply in general terms. To purists' objections that I may have cut too many corners, my answer is that refinement in understanding must be balanced between practical capability and a theoretical ideal. Refinements to knowledge are best added to improve finesse, but a basic understanding is the first essential. Similarly I have tried to show how many apparently complex interactions in the living plant are manifestations of the strong attractive forces between molecules, especially water molecules. Phenomena of membrane transport, gaseous transport and osmotic relationships have been drawn together in a simple mechanistic way to show how they interrelate. An integrated picture of plant water relations has seemed to me too difficult to derive from the many texts available and I hope this book will improve the situation.

I hope that the text is neither too simplified nor over-theoretical. Advances in understanding are often made by empirical methods, which may be more useful than theoretical treatments which may become too unwieldy. In this book I have paid considerable attention to the structural features of the systems studied; no self-respecting physiologist dares to neglect plant anatomy or ultrastructure. When I have had to choose between well-known material and neglected data, I have not hesitated to choose the latter, especially if I have found the experience personally rewarding, in the hope that some of the fascination reaches the reader.

We must remember that an understanding of plant water relations has wide implications in many disciplines, ranging from more efficient production of crops to the capacity to control our environment and increase protection from adverse influences. Our very survival is so tightly bound to plant growth that it is folly not to try to understand it as best we can.

John A. Milburn
Glasgow, 1977

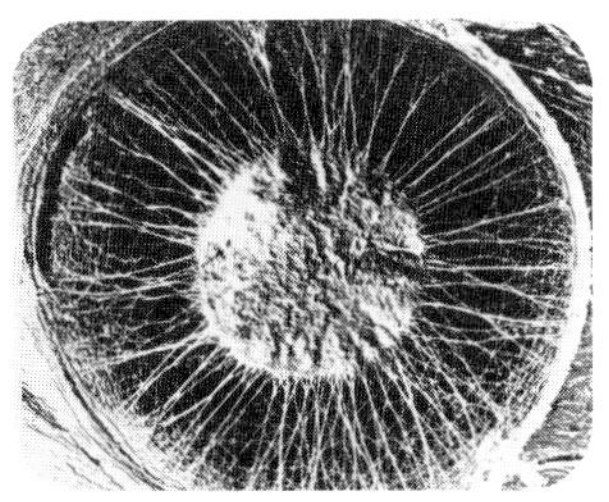

Acknowledgements

I am happy to acknowledge help from many colleagues. Dr Janet S. Sprent gave permission to use our scanning electron micrographs and Dr G. Puritsch and Dr R. P. C. Johnson provided the cover design micrograph. Dr D. A. Baker, G. Trevelyan and Dr D. Weeks provided useful criticism of the text. Drs B. Levine, T. A. Mansfield, R. C. Cogdell and L. Jones gave advice on specific points. Help with unpublished experiments and illustrations was provided by W. Perrie, B.Sc., A. McLellan and Miss Christine A. Lindsay, M.A.; Mrs C. Grant and Mrs C. Parker assisted with the script.

Dedicated to family and friends who suffered neglect during the preparation of this book.

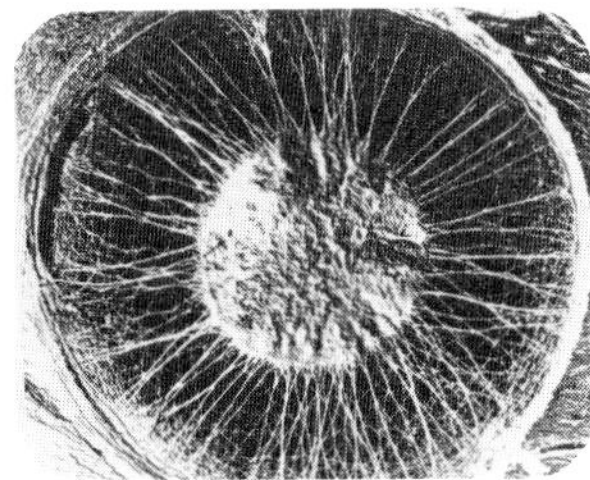

Chapter 1

Properties of water

Most living plant material is highly hydrated with a water content ranging from 70 to 90 per cent of the fresh weight. But this water is not stationary: it is constantly on the move. Water evaporates into the atmosphere and is thus drawn through the tissues of a plant from the soil. Plant material can lose many times its total water content per hour in the form of water vapour. *En route* through the plant tissues this water is subjected to negative pressure (i.e. tension) in xylem conduits and cell walls, yet each time water enters a living cell it is normally subjected to positive pressure. As sap tensions increase in response to environmental changes, water ebbs from the living cells, only to flow back again when the tensions are reduced. These fluctuations take place by the minutes, the hours, the days and the season. How does this amazing system work so silently and efficiently? Our understanding of the mechanisms involved must be based on the basic properties of water which are exploited most effectively by plants.

Despite its common occurrence, water is a highly unusual substance. Since it is a hydride of a non-metal it can be compared with other hydrides. Most of these, however, are not liquid but gaseous (e.g. NH_3, H_2S). Consequently, low temperatures or very high pressures are needed to liquefy many hydrides. Yet water is liquid at the relatively high temperature range of 0°–100°C and freezes readily to form ice. This phenomenon depends on the structure of the water molecule. Most hydride molecules involve strong chemical bonds within the molecule but only weak van der Waals' forces between the molecules. Water molecules are also held by strong internal covalent bonds with dissociation energy (the energy needed to pull the atoms apart) of 20 kJ mol^{-1}. Reflecting this strong bonding, the distance between hydrogen and oxygen atoms is only 0.099 nm. In contrast, the van der Waals' bonds have a very weak dissociation energy of about 1 kJ mol^{-1}.

The force which causes water molecules to aggregate to an extent abnormal in other hydrides is the hydrogen bond. This has an energy of dissociation midway between van der Waals' bonds and the virtually permanent covalent bonds. When two hydrogen atoms combine with an oxygen atom a 'lone pair' of electrons remains in the electron shell of the oxygen atom which is therefore locally electronegative (Fig. 1.1*a*). This localised charge distribution polarises the whole molecule so that it behaves like a minute magnet (i.e. a dipole): the hydrogen atoms to the exterior are electropositively charged. Hydrogen bonds are formed between oxygen atoms of one molecule with the hydrogen atoms of the next forming $(H_2O)_2$ in vapour. In ice irregular, roughly hexagonal and pentagonal bonding occurs (Fig. 1.1*b*). It will be noted that in each molecule from a sample of pure water the angle between covalent bonds of 105° imposes a three-dimensional structure.

Liquid water is imagined to resemble the structure of ice, except that hydrogen bonding is more ephemeral, being rapidly made and broken between molecules. At any instant of time the statistical chance of bonds being formed depends on the temperature.

About 15 per cent of the bonds are broken when ice melts at 0°C and the number of bonds broken rises to around 30 per cent of the total at 100°C: most biological systems depend on reactions in an aqueous medium within the temperature range of 0°–45°C (Fig. 1.1*c*). Hydrogen bonding can explain not merely the liquidity of water but other important properties including adhesion, cohesion, surface tension and viscosity which are discussed later. In response to temperature or pressure changes the bonding frequency per unit of time also changes, causing the water molecules to oscillate together in different patterns and their light absorbance to change. Accordingly bonding can be studied spectroscopically. These properties and notes on their physical and biological implications are listed in the Appendices (pp. 187–207). Water molecules also tend to adhere to solid surfaces, such as glass; this attraction can also be explained by hydrogen bonding. Large molecules, especially proteins like gelatine, also attract water; gels are formed and the large molecules are invested in massive hydration shells of water. Similarly smaller molecules and ions form hydration shells, indeed this explains a puzzle because one might expect, on the basis of relative atomic mass, the sodium ion (rel. at. mass 22.9) to diffuse in aqueous solution more rapidly than the more massive potassium ion (rel. at. mass 39.1).

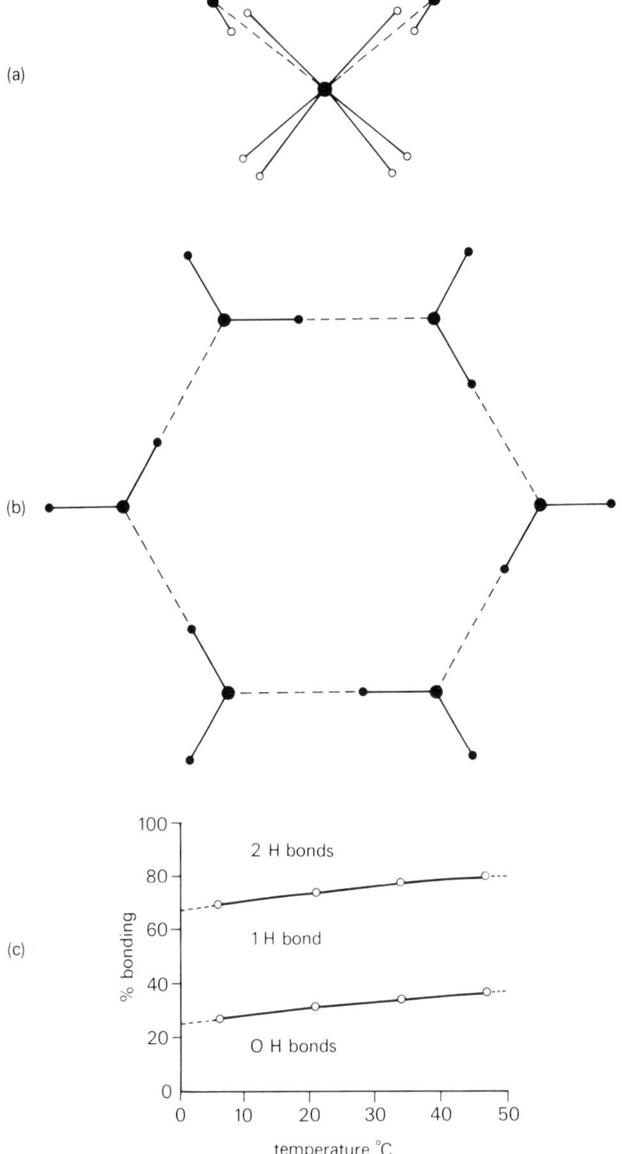

Fig. 1.1 (*a*) Diagrammatic representation of a water molecule to illustrate how electrons (○) are shared between hydrogen atoms (●) above, and an oxygen atom (●) below, to leave two lone pairs of electrons. (In terms of electron density the water molecule is 'V shaped' and 'fatter' than shown.) (*b*) Illustrations of the way molecules of water might attempt to link hexagonally through hydrogen bonding (dotted lines). For a regular hexagon the angles would be 120° but water is 105° preventing a regular flat configuration. (*c*) Showing the effect of temperature on bonding between water molecules. As the temperature increases the percentage of broken bonds increases (from Buijs and Choppin, 1963).

In fact potassium ions diffuse more rapidly; sodium ions are hindered by their larger hydration shells.

If sodium chloride crystals are added to ice, at say $-5°C$, all the crystals dissolve, producing a solution of brine, because the attraction between sodium and chloride ions for water exceeds that between water molecules in ice. Similarly when water molecules evaporate forming gaseous water vapour, an attraction also remains between the liquid and its gas molecules, the magnitude of which eventually controls the saturation level of water vapour, i.e. the amount of water held at equilibrium in the gas phase at a given temperature.

Thermodynamic terminology

The above outline of the molecular behaviour of water differs somewhat from that now in widespread use based on thermodynamic principles. This approach, expounded in innumerable texts, such as Slatyer (1967), is based on the energy status of water molecules, technically called the partial molar *Gibbs' free energy* or *water potential* (see App. 17). According to this approach an aqueous specimen in which the free energy has been modified by changes in pressure, or the addition of solutes is compared with an arbitrary standard. The *water potential* Ψ (psi) of this standard, which is pure water at atmospheric pressure and the same temperature as the unknown specimen, is arbitrarily set at zero energy units.

There are several difficulties with the thermodynamic approach. The concept itself is difficult to picture. Though the units of measurement are ideally based on the energy per mole ($J\,mol^{-1}$), in practice they are considered in terms of energy per unit volume ($J\,m^{-3}$ from the partial molar volume). Though the difference is small, it is confusing, as pointed out by Spanner (1973). Because the arbitrary standard is zero, water potentials are positive or negative in plants which makes calculation tricky. Finally, measurements of water potentials are made with instruments calibrated in pressure units, such as pressure bombs, manometers, tensiometers and osmometers. Nevertheless water potential terminology is interdisciplinary and comprehensive. To meet these requirements, an alternative *concept*, based on pressure, is outlined below. This is not a serious departure from thermodynamic definitions and water potential terminology is retained, but our approach will be based on pressure which is the practical unit of measurement.

According to this approach the tendency of water molecules to cohere is conceived as an *internal pressure* (see Moore, 1963). The internal pressure is the resultant of the forces of attraction and the forces of repulsion between the molecules in a liquid. In liquid water the molecules are held together by hydrogen bonds, as if by suction. If the molecules become more separated hydrogen bonds cannot form so readily and the internal pressure is reduced (suction increases). This basic idea allows water to be considered as a 'tensile solvent', see below. Plants exploit the tensile properties of water to a remarkable extent.

Water potential Ψ, internal pressure and definition

When we consider liquid water rather than water molecules, we become less interested in the very rapid changes in the energy of individual molecules and more concerned with the mean condition of whole populations of molecules. The fact that water is normally liquid rather than a gas is a consequence of its cohesive properties caused by hydrogen bonding which holds water molecules together by powerful internal tensile forces. Cohesion can be imagined as powerful internal pressure between water molecules making water a 'tensile solvent', a concept developed with some rigour by Hammel and Scholander (1976).

As liquid water is heated above boiling point (100°C at atmospheric pressure) the cohesive bonds in the liquid are broken so that it becomes gaseous. *Positive* pressures must be applied to restore the liquid state. Conversely liquid water can be made to boil at ambient temperature by subjecting it to strong negative pressures which we may call tensions to dispense with the negative sign. These tensions are very difficult to measure experimentally but theoretical estimates of the pull required, which is the tensile strength of pure water at ambient temperatures, lie between −1,000 and −20,000 bar (see also App. 5). Since this pressure cannot be measured accurately, we make pure free water at atmospheric pressure our *arbitrary* reference standard. Such water on the above scheme has a *water potential* Ψ of zero internal pressure units (we use the bar and the pascal in this book). We may observe that had we used energy units, say $J\,m^{-3}$, the dimensions would be energy per unit volume ($ML^{-1}T^{-2}$).

Water potential Ψ, by our internal pressure definition is the difference in intramolecular pressure exerted in a given specimen with reference to the intramolecular pressure exerted between molecules of pure free water at atmospheric pressure and the same temperature. The water potential of pure water is therefore raised by the application of external positive pressure or lowered by the application of negative pressure. At a molecular level we can imagine that water molecules being dipolar, become more nearly aligned by negative pressure, like strings of small magnets, so reducing the freedom of hydrogen bonding. Conversely positive pressure increases the capacity of water molecules to form hydrogen bonds so raising their water potential which represents their chemical reactivity. Changes in temperature affect water potential in a manner similar to pressure: these are discussed in Chapter 6.

The presence of foreign molecules such as solutes, gases or molecules attached to surfaces reduces the reactivity of water in exactly the same way as external negative pressure. It is convenient to separate the components which make up the composite water potential Ψ and they will be considered in turn below.

Pressure potential Ψ_p

The component of Ψ influenced by pressure will be designated Ψ_p. Water in plants is commonly subject to suction or tension (also called negative pressures and tensile stresses) which reduce Ψ_p below zero making Ψ_p, when a 'suction', negative in sign. Since our reference standard is pure free water at atmospheric pressure and the same temperature then Ψ_p of liquid water under vacuum is -1.013 bar. (It should be noted that this contrasts with the convention in gas law studies by which vacuum represents 'absolute zero' of 0.0 bar.) In plants Ψ_p can approach ≈ -100 bar!

Solute potential Ψ_s

Foreign molecules dissolved in water attract some water molecules thus *always* increasing the attraction between the remaining water molecules. Water potentials so modified are always negative in sign and are symbolised Ψ_s, the component of water potential influenced by *solutes.* Theoretically foreign molecules might repel water molecules and so produce positive water potentials raising Ψ_s above zero: in practice such molecules are insoluble. (A most interesting exception to this rule is caused by small hydrocarbon molecules which are hydrophobic. Their presence causes water to 'freeze' at temperatures above 0°C forming clathrates. This effect is used to purify water commercially and may explain why some crops such as maize are sensitive to frost damage at temperatures above 0°C.) Thus Ψ_s is called the *solute or osmotic potential* of a solution and is practically *always a negative* water potential. If a water potential is entirely influenced by solute molecules $\Psi = \Psi_s$ in both magnitude and sign. Both are termed potentials. While there is no way to measure the mean negative pressures (tension) between water molecules directly, it *is* possible to measure the differences in mean internal pressure between molecules in a solution and those in pure water by enclosing the solution in an osmometer which is an instrument in which a solution is separated from water by a membrane called a semi-permeable membrane permeable only to water (the solvent) but not the solute molecules (see Ch. 2). Free water outside an osmometer is drawn through the semi-permeable membrane by the greater intramolecular tensions within (see Fig. 1.2), a process called *osmosis.*

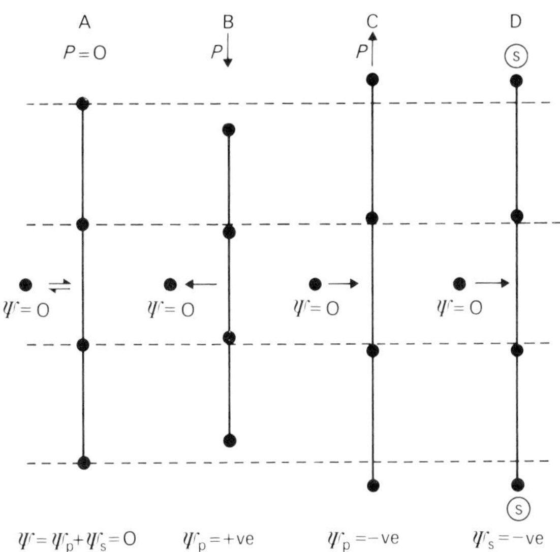

Fig. 1.2 Diagrammatic representation of the water potential of pure water A, and water under the influence of a positive pressure B, which compresses the molecules; or a negative pressure C which 'stretches' the water molecules. A reference water molecule ($\Psi=0$ see central positions) may be attracted or repelled by the mass of water molecules. The addition of solute molecules D attracts shells of water molecules 'stretching' the remaining water molecules in the same way as negative pressure in C.

Eventually equilibrium is established at maximum internal volume and a positive hydrostatic pressure within the osmometer counteracts further net water entry. This positive pressure is called the *osmotic pressure.* It is identical numerically but opposite in sign to the solute (or osmotic) potential Ψ_s. Thus at equilibrium throughout the system, inside and outside the osmometer, the water potential $\Psi=0$ bar which equals the $\Psi_p+\Psi_s$ components within the osmometer. Accordingly if Ψ_p, the osmotic pressure of a solution, is 5 bar then Ψ_s (its osmotic potential) must be -5 bar. Consequently Ψ_p, when caused by Ψ_s in the presence of pure water, is the osmotic pressure (for convenience it can be given a separate symbol Π, pi). Some workers prefer to use an additional component Ψ_m, the matric potential which according to our system would be added to Ψ_p and

Ψ_s. Soil scientists use this term to study water-holding by particles by surface tension or surface (imbibitional or hygroscopic) forces. There seems little point in so doing however, because surface tension effects can rightly be included with the Ψ_p component. On the other hand imbibition depends on the attraction of water to surfaces which may be rigid, colloidal, or molecular as in the case of gelatine molecules, but it does not differ significantly from Ψ_s in principle and may be grouped with this component. Similarly a gravitational component Ψ_g is occasionally used for hydrostatic water potentials but this is really a part of Ψ_p component. There is no objection in principle to using any such terms for convenience, but Ψ_p and Ψ_s are simple and adequate for most purposes (see also App. 17).

The measurement of water potential Ψ

In any situation the net water potential Ψ is the sum of hydrostatic pressure Ψ_p and solute Ψ_s potentials

$$\Psi = \Psi_p + \Psi_s \qquad [1.1]$$

In land plants under natural conditions Ψ is negative but it can become weakly positive (e.g. during guttation). Methods can be chosen to measure Ψ_p and Ψ_s together or separately. Changes in Ψ represent changes in the extent of hydrogen bonding which also modifies the colligative (glue-like, from Greek: *collos*) properties of water. Thus if Ψ of a solution is reduced its freezing point and vapour pressure are depressed while the boiling point is raised. Changes such as these are exploited to measure the more difficult negative water potentials. Perhaps the simplest measurements to make in principle are positive water potentials.

Pressure potential Ψ_p

Positive water potentials above our reference standard of free water at atmospheric pressure, can be measured with great ease using a simple manometer. The difference in height between liquids in the limbs in a U tube registers pressure difference Ψ_p ($=h\rho g$) relative to ambient, i.e. atmospheric, pressure. Other designs of pressure gauge (Bourdon and bubble types) are described later.

Small negative hydrostatic pressures down to -1 bar can be measured manometrically also, but at more extreme negative pressures manometers become so unstable through cavitation as to be useless. Negative hydrostatic pressure produces the same reduction in the water potential of cells as is caused by the addition of solute molecules and can be measured by this means. Sap tensions below

-1 bar, which are common in xylem, can also be measured using a pressure bomb (see Ch. 5).

Water potentiometer

In theory a negative water potential at a point can be measured in location very simply by allowing a small drop of osmotic liquid to equilibrate in the immediate vicinity. If the water potential of the system is -2 bar and a -1 bar droplet is allowed to equilibrate, it will lose water by distillation until its volume is halved and its osmotic potential has doubled. In this system air acts as a perfect osmotic membrane. Equilibration is often very slow and must occur under isothermal conditions. These requirements make the techniques difficult to apply in many instances. (A small correction is required also from surface tension acting in the droplet.) If the droplet is placed in direct contact with a tissue or cell surrounded by a semi-permeable membrane, equilibration is rapid and less dependent on temperature.

Osmometry

The solute potential Ψ_s of the droplet equals the water potential of our unknown system resulting both from hydrostatic and solute potentials (providing the net water potential Ψ is negative). Theoretically Ψ_s of the small droplet could be determined in an osmometer but this is not practical and other means are more convenient. We might, for example, use a '*freezing-point osmometer*' which registers the depression of freezing point of our sample below that of pure water. Such an osmometer usually detects the point of crystallisation from the latent heat of fusion of ice on a sensitive electrical thermometer. Alternatively the temperature at which ice crystals just dissolve in a tiny sample, observed under the microscope, may be measured (Ramsay's method).

$$-\Psi_s = \Pi = 12.06\Delta T - 0.021\ (\Delta T)^2 \text{ (bar at 0°C)} \qquad [1.2]$$

Where ΔT is the depression of freezing point (see Crafts *et al.*, 1949).

Psychrometry

Another method assays water potentials by measuring the temperature reduction resulting from the heat of vaporisation of pure water compared with a specimen. Psychrometers embody highly sensitive thermometers and can only measure *negative* water potentials, which in plant materials may comprise both pressure and solute components. In practice psychrometers are usually calibrated against liquids of known osmotic potentials such as sucrose or sodium chloride solutions. Clearly a psychrometer can be used to measure water potentials directly on plant material, or on droplets of solution (where it is called a '*vapour-pressure osmometer*'), or even the atmospheric humidity (as a *hygrometer*). Water potential, Ψ, is related to atmospheric humidity by the expression

$$\Psi = \Psi_s = -\frac{RT}{\bar{V}} \ln\left(\frac{\% r.h.}{100}\right) \quad \text{(bar at } T\text{K)} \qquad [1.3]$$

where Ψ_s is the solute potential in the droplet, R is the gas constant, T is the temperature, r.h. is the relative humidity, and $\bar{V}$ is the partial molal volume of water.

Other methods – refractometry

Several other methods can be used to measure the osmotic potential of droplets, providing the composition of the liquid is approximately or exactly known. If a pure known solute is used, such as sodium chloride, an electrical conductivity meter may provide a convenient measure of its concentration. Another extremely convenient method is to use a *refractometer*. Concentrations of single droplets can be determined in a few seconds. The refraction of solutions provides a precise and almost perfectly linear relationship against molal concentration (see App. 12). It is relatively easy to produce osmotic potential – refractive index calibration curves. If plant saps are measured the method is less precise, owing to the fact that the refraction differs between solutes and so varies with sap composition. The extent of this error can be assessed by plotting a calibrated regression curve.

Other methods are based on differences in density such as the Schardakov method. Replicate samples of plant material are immersed in a range of solutions of different Ψ_s values, containing a small amount of dye. After an equilibration period (about 30 min) droplets of the solutions are gently injected into glass vials of the same solutions taken originally without dye. If the plant material has extracted water, the dyed droplets sink; if the reverse, the droplets rise owing to a reduction in their density. The water potential of the solution droplets which neither rise nor sink equals the water potential of the plant tissue.

For a fuller review of methods see Barrs (1969).

Description of water transport systems

This section outlines the basic mechanisms which drive water flow. A later section examines some important quantitative relationships.

Mass flow

The simplest method to drive liquid water or solution flow is by externally applied pressure. Mass flow (also called bulk or pressure flow) implies a movement of molecules in unison – nevertheless molecular diffusion occurs within the moving stream so that some molecules move faster, others slower than the average speed. Mass flow can be represented in the tube in Fig. 1.3. Flow is driven by a piston, but the pressure gradient could alternatively be driven by other means such as a gravitational head of liquid. Flow along the tube is modified by frictional drag along the walls. The behaviour of water is 'ideal' in this respect, with a paraboloid flow pattern. Such a

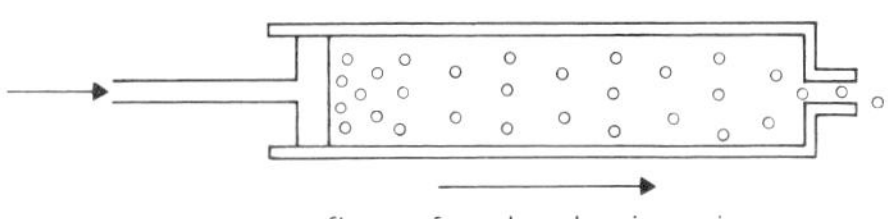

Fig. 1.3. Diagrammatic representation of mass or bulk flow of liquid water in a syringe. A wave of compression is initiated by the plunger and travels at the speed of sound in water (1,498 ms^{-1} at 25°C).

pattern is termed Newtonian in contrast with non-Newtonian systems such as flows of pastes, slurries, or viscous fluids. Another interesting point is that if the cross-section of the tube is reduced, say by half, the flow rate tends to double if the pressure difference is maintained, but the pressure on the walls of the narrower tube is not increased, as one might expect, but decreased. This effect is called the Bernouilli principle and is of importance in xylem transport (see Ch. 5).

Pressure drives flow from a region in which water molecules are more compressed (therefore more concentrated, with a higher internal pressure and with greater hydrogen bonding) towards regions in which water molecules are less compressed (therefore more diluted, with a lower internal pressure and more free to form hydrogen bonds). In fact liquid water is almost incompressible so the difference in compression is very slight, but it is highly significant nevertheless. The time for a pressure wave to travel in the liquid is very short (the velocity of sound in water, see App. 2) but it too is significant when considering the propagation of flow. Since mass flow is driven from more compressed to less compressed liquid, flow can be regarded as moving from high to low concentrations. In this way mass flow can be compared with diffusional flow.

Diffusional flow

Water molecules are in a constant state of molecular agitation which causes localised variations in pressure within the liquid. These induce random mixing of molecules, a process called *diffusion*. In pure water there is no net drift of molecules in any particular direction. If solute molecules, say in the form of a droplet of dye, are gently injected into water, the droplet gradually seems to enlarge and disperse. This is because of a net drift of water molecules flow into the dye droplet along an internal pressure gradient. The internal pressure of the solution is effectively reduced below that of water through hydrogen bonding between water and solute molecules. This flow, called *diffusional flow*, takes place even though at a macroscopic level the hydrostatic pressure is uniform throughout both water and solution. As in the case of mass flow, water seems to flow from high concentration of water to lower concentrations of water and the same applies to the reverse movement of solute molecules (Fig. 1.4).

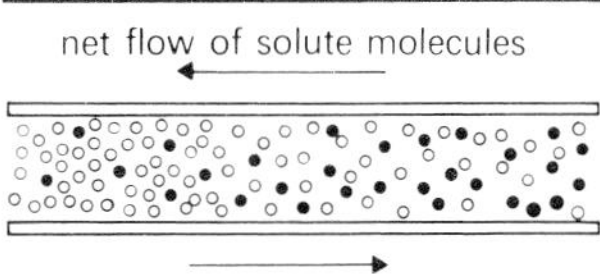

Fig. 1.4 Representation of diffusional flow of water (○) and solute (●) molecules in a tube as counter currents driven by concentration gradients.

The diffusional transport process is very slow over long distances, while diffusional flow over short distances is rapid, an important factor controlling water flow and solute uptake by roots and solute movement between cells. For practical purposes diffusion constants (*D*) are used to calculate the rates of diffusion of different solutes and also the different isotopes of water (D_2O, T_2O) in water. There is little difference in principle if water molecules are separated by solute molecules or, in the form of vapour, by gas molecules, though there are different diffusion coefficients for the liquid and gaseous states. Gaseous diffusion flow is important during gas exchange by leaves when water molecules become widely separated as they vaporise.

Osmotic flow

If a selective barrier, permeable to water but not solute molecules, is situated where there might otherwise be mutual diffusional flow between solutions (Fig. 1.5), a special phenomenon occurs called

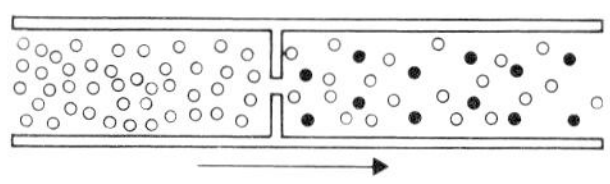

Fig. 1.5 Osmotic flow across a semi-permeable membrane with a single pore. Water molecules (○) penetrate the pore under a local pressure gradient until opposed by a general pressure gradient but solute molecules (●) cannot escape.

osmosis. Water moves from one side of the barrier where it has a higher internal pressure (it is more concentrated) to the other solutes with a lower internal pressure (it is less concentrated). Thus a net flow of water is established which seems unidirectional because the larger solute molecules are unable to penetrate the pores of the molecular sieve, called a *semi-permeable membrane*.

The flow of water molecules through pores in the semi-permeable membrane is of course mass flow driven by pressure, but immediately they reach the solute molecules, movement is by diffusion, which tends to limit the rate overall. Thus the *local pressure*, on the A side of the pore with pure water, is zero, but as soon as a solution is introduced into B, a localised negative pressure is generated by a reduction in hydrogen bonding between water molecules within the solution in B adjacent to each pore, pulling water through it. Since the flow rate increases if the concentration of solute molecules in B is increased, the inaccurate concept was originally developed of a suction of water by the solute molecules; hence the older term suction force (SF). Later, as the diffusional nature of the process involving water rather than solute molecules was better understood, the term was replaced by diffusion pressure deficit (DPD) to describe the state of the water. These terms are *numerically identical* with water potential but *opposite in sign*. Water potential Ψ terminology is now both standard and interdisciplinary so will be used throughout this book.

Liquid to vapour transport

Plants evaporate very large amounts of water by transpiration, which is the transfer of liquid water into the vapour phase (Fig. 1.6). In many respects transpiration resembles the osmotic processes described above.

When water molecules are vaporised they become too widely separated for hydrogen bonding to be very significant. The water potential of a vapour is mainly affected by the concentration of the water molecules which have free hydrogen bonding sites (lone pairs of electrons). The greater the mean distance between the water molecules the lower is the water potential. Of course if the concentration

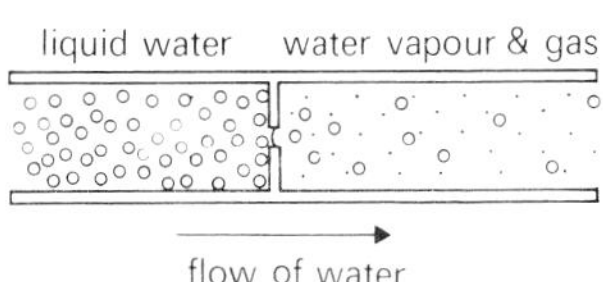

Fig. 1.6 Escape of liquid water as vapour into a gas across a film (across a pore) representing surface tension. In the gas the flow is diffusional.

exceeds a certain limit (saturation) vaporised water molecules condense to the liquid state. When condensation occurs hydrogen bonds are reformed and energy is released as heat. Conversely the evaporation of water requires additional energy to break hydrogen bonds – the latent heat of vaporisation, which cools the liquid surface during evaporation. In practice the heat supply often limits the rate of evaporation or transpiration from leaves.

Evaporation is also strongly influenced by the temperature of the vapour which regulates the thermal agitation of the molecules and hence the maximum concentrations possible. An increase in temperature lowers the water potential of a sample of vapour. Evaporation is affected similarly by the gaseous pressure but a *decrease* in pressure *increases* the diffusion rate of water molecules through a gas because interaction with gas molecules (internal friction) is decreased. (It is for this reason that gas pressure is reduced to near vacuum when freeze-drying a specimen – the rate of evaporation of water is greatly increased.) In the simple model shown in Fig. 1.6 evaporation occurs from A in which liquid water is held by surface attractive forces of the container, its own cohesion and surface tension into B. The number of water molecules per unit volume is much greater at A than B, but according to our model the extent of vacant hydrogen bonding sites will be approximately uniform throughout when equilibrium is reached.

At A *the number of molecules per mol is expressed by Avogadro's number* N *so that 1 mol of water, weighing* 18.048×10^{-3} kg at 20°C *contains* 6.023×10^{23} *molecules, or* 3.3×10^{22} *molecules* ml^{-1}. *In contrast in* B, *which we will assume contains water-saturated air at* 20°C, 17.3×10^{-9} *molecules of water* ml^{-1} *air which equals* 5.76×10^{17} *molecules, a much smaller number. Nevertheless it will be appreciated that both numbers are extremely large. For this reason we can talk about a pressure as a constant figure, although pressure represents the random movement of individual molecules. Each molecule is statistically insignificant with respect to the whole population sampled. We cannot detect pressure variations unless we examine small particles microscopically. Their erratic 'dance' arising from an imbalance in molecular collisions causes 'Brownian motion' named after the botanist, Robert Brown, who first described it.*

Interrelation of pressure, vapour and osmotic systems

It is possible to set up a combination of some of the systems we have already examined to illustrate the interrelation of water potential Ψ_p to Ψ_s in the liquid and vapour phases. These interrelations affect all land plants.

The system shown in Fig. 1.7 consists of water held in a porous

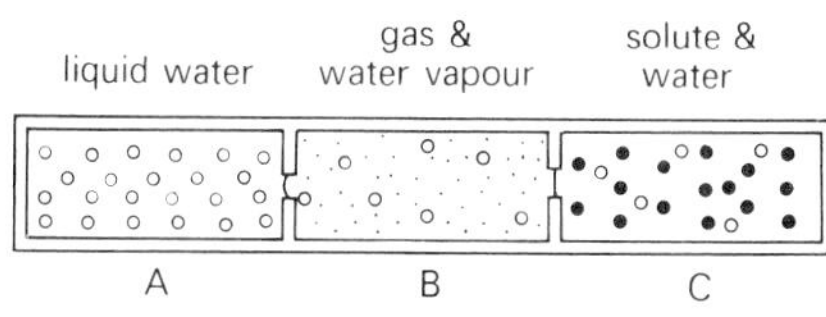

Fig. 1.7 Diagram showing a sealed vessel containing three compartments at equilibrium. In each the water potential is identical thanks to negative pressure in A, gas in B and solute molecules in C but there is considerable disparity in the actual concentration of water molecules.

matrix of cellulose or wood at A so that its volume is effectively fixed. At B the gas phase is partly saturated with water vapour. Some solution, of $\Psi_s = -5$ bar (which in an osmometer could generate a positive osmotic pressure, Ψ_p, of 5 bar) is sealed in C. At equilibrium water potentials equalise throughout A, B and C. Since A contains pure water it must assume the same water potential as C by developing a negative pressure acting on the walls so that Ψ_p is -5 bar. This is possible only if the pore between A and B is sufficiently narrow for the surface tensional forces of water molecules to operate an effective pressure seal. We will presume this. At equilibrium the measurable pressure in A is therefore -5 bar lower than normal atmospheric pressure (our reference pressure 0.0 bar) which applies in B and C.

The gas contained in B is interesting for three reasons. First, it may be noted that while equilibrium is being established it operates like a perfect semi-permeable membrane allowing the movement of water molecules yet preventing the escape of solute molecules from C. Unlike most membranes, flow through the gas phase is entirely by diffusion and not mass flow. Second, we may observe that though B is at ambient hydrostatic pressure like C, the water potential is -5 bar reflecting a reduction of the concentration of water molecules in the gas below saturation. In fact this is very slight; the relative humidity of the water vapour would be 99.7 per cent, very close to the saturation point of 100 per cent at 20°C. Third, it is important to note that even a slight temperature gradient between A and C, causes a considerable change in the equilibrium (1°C is equivalent to $\Delta\Psi$ of 81.6 bar, or 1 bar$\equiv\Delta 0.012$°C).

Electrically driven flow

To a very slight extent water dissociates into ions, protons (H^+, or more strictly H_3^+O) and hydroxyl ions (OH^-), the concentrations of which reflect the acidity or alkalinity of a medium; hence pH is a derived scale as a log concentration of the H^+ scale to allow the use of small numbers (pH $= -\log_{10}[H^+]$ so that pH 1 is 10^{-1}, pH 2 is 10^{-2} protons as mol l^{-1}). Hydrogen ion concentration, which is greatly influenced by the presence of other solutes, has long been known to have a profound effect on living processes and also the activity of complex molecules such as enzymes and other proteins. Recently it has become apparent that cell membranes control electrical gradients through proton extrusion. Since the movement of any ion is influenced by both electrical and concentration gradients this promises to be an important avenue of new research (see Ch. 4).

Each water molecule is itself a dipole with weak, but significant, 'magnetic' properties. Thus when water molecules flow they impede a counter current of electrons, an effect which is exploited to measure liquid flow. Similarly a flow of water influences an imposed magnetic field, which provides the principle of the hydrodynamic flowmeter (see Ch. 5 for more details of both systems).

Furthermore the flow of water molecules over a surface induces an electrical potential gradient. Conversely if a porous medium is electrically charged it is possible to drive a current of water or produce a hydrostatic pressure: this principle is called *electro-osmosis*. As yet electro-osmotic flow has not been proved to occur in plants, but it may well develop during sap flow through porous systems, e.g. cell walls, and some have proposed that this mechanism contributes to sap transport in the phloem.

Equations describing flow quantitatively

Many problems in plant physiology can only be resolved by examining them in detail and ultimately quantitatively. Many flow equations describing water flow in plants are analogous to Ohm's law. Since this is simple and well known it will be used as our starting point.

Electrical flow

Many flow equations are closely analogous to Ohm's law which describes the flow of electrons and which we will use as a model. Flow is described in terms of the driving force represented by a voltage differential E and the resistance to flow R, measured in ohms, according to the simpler relationship

$$\text{Ohm's law} \quad I=\frac{E}{R} \text{ and, since } \frac{I}{R}=C \text{ so } I=CE \qquad [1.4]$$

where I represents the flow or current in amperes, E the voltage difference in volts, R the resistance in ohms and C the conductance in siemens (reciprocal ohms). Electronic definitions tend not, however, to be readily interpretable in quantities of electrons, probably because it is much more difficult to handle electrons, compared with, say, a given volume of water.

Considerable confusion has occurred in the past owing to differences in terms, symbols, dimensions and units applicable to equations such as those outlined below. For this reason below each major equation presented here is a list of terms, symbols, dimensions and units. The importance of dimensional checks involving algebraic cancellations to check equations cannot be overemphasised. Occasionally it is useful not to reduce dimensions to basic units but to leave them in, for example, pressure units (Pa or bar) not $\mathrm{kg\,m^{-1}s^{-2}}$. In this treatment the derivation of equations is omitted: each equation being regarded as a 'black box', ready to be used by plugging in components (under manufacturer's instructions) to answer questions quantitatively.

Hydraulic flow equations

Probably the most basic equation describes the conductivity of a system in terms of the volume or hydraulic flux driven by a difference in water potential. It is closely analogous to Ohm's law.

$$J_v = L_p \Delta\Psi_p \qquad [1.5]$$

Hydraulic flux	J_v	$L^3L^{-2}T^{-1}$	$\mathrm{m^3\,m^{-2}\,s^{-1}}$ ($=\mathrm{m\,s^{-1}}$)
Hydraulic conductance	L_p	$L(ML^{-1}T^{-2})^{-1}T^{-1}$	$\mathrm{m\,Pa^{-1}\,s^{-1}}$ ($=\mathrm{m\,bar^{-1}\,s^{-1}}$ $\times 10^{-5}$)
Difference in water potential	$\Delta\Psi_p$	$ML^{-1}T^{-2}$	$\mathrm{kg\,m^{-1}\,s^{-2}}$ ($=$Pa)

It is important to realise that L_p is a measure of conductance of a specimen of wood etc. irrespective of its length. For comparative purposes it is often more useful to use a different coefficient L, which includes the length of the flow pathway. This equation departs from Ohm's law, and L is the main comparative unit for hydraulic conductivity used in this book. Parameters additional to those above are given below:

$$J_v = L\frac{\Delta\Psi_p}{x} \qquad [1.6]$$

where $J_v = \dfrac{V}{At}$ and $L_p = \dfrac{L}{x}$

Hydraulic conductivity	L	$L^2(ML^{-1}T^{-2})^{-1}T^{-1}$	$m^2 (Pa)^{-1} s^{-1}$ ($= m^2\ bar^{-1}\ s^{-1} \times 10^{-5}$)
Volume transported	V	L^3	m^3
Area of cross-section	A	L^2	m_2
Time of flow	t	T	s
Length of path	x	L	m

Water flow through capillaries and pores

In 1839 and 1840 Hagen and Poiseuille independently developed a most useful relationship to explain in quantitative terms the capacity

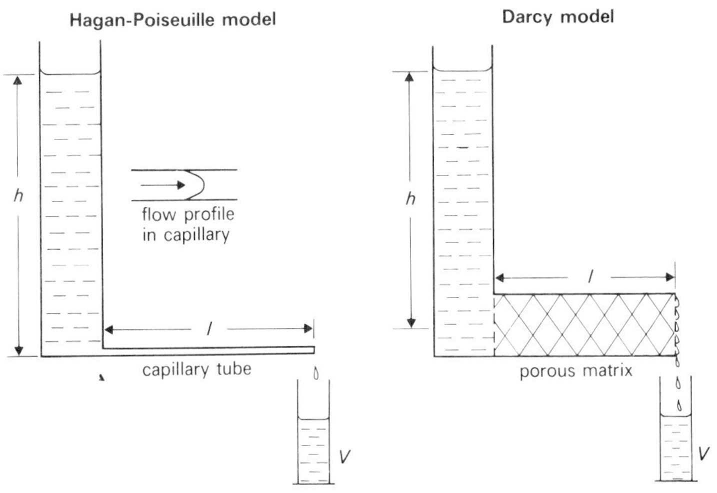

Fig. 1.8 Two models to illustrate the Hagen-Poiseuille system of flow (left) through a capillary tube and Darcy flow through a porous medium (right). In both models a pressure difference is generated by a manometric head of water, h. The flow rate across a specimen path, length l, is found by timing the collection of volumes of water. *Note:* $l = x$ in equations [1.6] and [1.7].

of smooth cylindrical capillaries to conduct liquids. Per unit cross-sectional area the flow through a smooth capillary is proportional to the square of the radius (see Fig. 1.8).

Hagen–Poiseuille Equation $$J_v = \frac{r^2}{8\eta}\frac{\Delta\Psi_p}{x}.$$

$$\text{Hence } L = \frac{r^2}{8\eta} \text{ in a capillary.} \qquad [1.7]$$

Hydraulic flux	I_v	$L^3L^{-2}T^{-1}$	$m^3 m^{-2} s^{-1}$)
Tube radius	r	L	m
Pressure potential difference	$\Delta\Psi_p$	$ML^{-1}T^{-2}$	$kg\,m^{-1}s^{-2}$ (=Pa=bar × 20^{-5})
Viscosity of liquid	η	$ML^{-1}T^{-1}$	$kg\,m^{-1}s^{-1}$
Length of path	x	L	m
Hydraulic conductivity	L	$L^2(ML^{-1}T^{-2})^{-1}T^{-1}$	$m^2\,bar^{-1}\,s^{-1} \times 10^5$

The flow of liquid through the tube is not uniform across its bore but paraboloid, like an extending telescope, with maximum velocity in the central axis of the tube. Friction of the walls prevents a sliding motion, so the outermost velocity tends towards zero. Consequently if we are interested in velocity movement of sap we must distinguish between mean velocity $\bar{u}$ and maximum velocity u_{max} of flow.

$$\mu_{max} = \frac{r^2}{4\eta} \cdot \frac{\Delta\Psi_p}{x} \quad \text{and} \quad \bar{\mu} = \frac{r^2}{8\eta} \cdot \frac{\Delta\Psi}{x} = J_v \qquad [1.8]$$

The equations above illustrate the remarkably simple relationship between u_{max} and $\bar{u}$ which differ only by a factor of 2 on account of the parabolic flow pattern. In other words, if u_{max} cannot be detected precisely (because the front tends towards zero concentration), it can be inferred from $\bar{u}$ or the shape of the profile. In most situations in plants the flow pattern is complicated because of the many tubes transporting in parallel. Because of the fourth-power effect of the radius wider tubes carry much greater volumes of fluid than narrower tubes (see Ch. 5).

Test for turbulent flow

Reynolds deduced a formula empirically in 1883 as follows to test for laminar or turbulent flow in capillaries.

Reynolds number $$\mathrm{Re} = \rho \frac{J_v 2r}{\eta} \qquad [1.9]$$

Reynolds number	Re	dimensionless	–
Liquid density	ρ	ML^{-3}	$kg\,m^{-3}$
Volume flux	J_v	$L^3L^{-2}T^{-1}(=LT^{-1})$	$m^{-3}m^{-2}s^{-1}$ (= $m\,s^{-1}$)
Tube radius	r	L	m
Viscosity (dynamic)	η	$ML^{-1}T^{-1}$	$kg\,m^{-1}s^{-1}$ (= Pa s)

It has been established experimentally that if Re is less than 2,000 the flux is laminar; but if greater, flow is turbulent. In most biological systems, even under exceptional circumstances, laminar flow is the general rule.

In 1856 Darcy deduced a law to explain the bulk flow of water through soil (Fig. 1.8) in modern form as follows:

Darcy's law $$J_v = L\frac{\Delta\Psi_p}{x}$$ [1.10]

Volume flux	J_v	$L^3L^{-2}T^{-1}$	$m^3\,m^{-2}\,s^{-1}$
Hydraulic conductivity	L	$L^2(ML^{-1}T^{-2})^{-1}T^{-1}$	$m^2\,bar^{-1}\,s^{-1}\times 10^5$
Pressure potential difference	$\Delta\Psi_p$	$ML^{-1}T^{-1}$	Pa (= bar × 10^{-5})
Path length	x	L	m

Darcy's law is used to deduce the hydraulic conductivity of any porous system including soils and cell walls.

Diffusive flow

Fick's first law was developed to explain the transport of molecules by diffusion. Despite the fact that it resembles Ohm's law in some respects, there are important differences.

$$J = D\frac{\Delta c}{x}$$ [1.11]

Molecular flux	J	$ML^{-2}T^{-1}$	$kg\,m^{-2}\,s^{-1}$
Diffusion coefficient	D	$L^3L^{-2}T^{-1}L(=L^2T^{-1})$	$m^2\,s^{-1}$
Concentration difference	Δc	ML^{-3}	$kg\,m^{-3}$
Path length	x	L	m

Note the difference in dimensions and units in comparison with previous equations. The driving gradient, being dependent on concentration, is relatively linear over short distances and free from temperature effects. For this reason Fick's law has been widely used to study flow through membranes, the escape of water during transpiration, and the processes of gas exchange generally. Diffusional processes can be extremely efficient providing x is short (see Chs. 3 and 6).

Over longer distances the time–distance relationship is not simple owing to the complexity of the diffusion profile itself (see Nobel, 1974). One useful approach is to measure the time required for a concentration 'front' to advance. The calculation is simpler if this concentration is set at 37 per cent of the value at the origin which may be maintained at a constant level or decrease as diffusion dilutes the initial concentration. The time for this front to move is given approximately by the following expression which is derived from Fick's second law of diffusion:

$$t = \frac{x^2}{4D} \qquad [1.12]$$

Time	t	T	s
Diffusion path (for 37%)	x	L	m
Diffusion coefficient	D	L^2T^{-1}	$m^2 s^{-1}$

Equation 1.12 shows that the rate of diffusional transport is approximately proportional to the square of the diffusion path length. For this reason diffusion is an important transport mechanism at the cellular level but quite ineffective for long distance transport in plants, as will become apparent in later chapters.

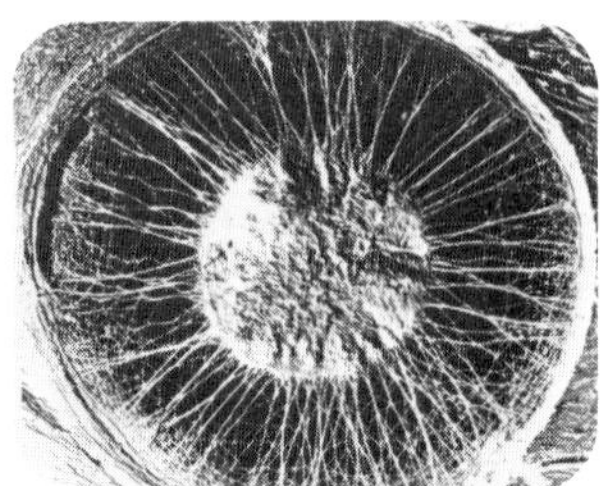

Chapter 2

Water machines and models

In this chapter we will explore some interesting water machines. The relatively simple physical principles on which they depend are merely an extension of the properties of water described in Chapter 1. Such machines or models provide a means to grapple with the basic principles operating in a given system. Plants operate by many interrelated systems which are too complex to be analysed directly, so a model is abstracted to illustrate a hypothesis embodying the basic principles supposedly operating in the system under investigation. When the analogy between a model and a plant has been established qualitatively the next stage of investigation is to check it quantitatively. Only if this test is successful can the hypothesis be considered proven. The following six machines and models illustrate the physicochemical basis for understanding water transport in plants. The first three examine pure water in a physical system and the remainder illustrate possible effects caused by the presence of dissolved solutes.

The water barometer

In any large natural body of fresh water the water potential at the surface is by our definition, 0 bar. At a depth of about 10 m the pressure component increases by 1 bar becoming +1 bar: continuing descent will increase the pressure to +2, +3 bar, etc. at intervals of about 10 m (see App. 4). This shows clearly how pressure can produce water potentials. The calculation of the pressure at 10.13 m is as follows:

$$P = \Psi_p = h\rho g = 10.13 \times 998.4 \times 9.81 \times 10^{-5} = 1.0 \text{ bar} \qquad [2.1]$$

Pressure *P* at depth caused by gravity	P, Ψ_p	$ML^{-1}T^{-2}$	Pa ($= 10^{-5}$ bar)
Height of column liquid	h	L	m
Density of liquid	ρ	ML^{-3}	kg m^{-3}
Gravitational constant	g	LT^{-2}	m s^{-2}

If an inverted glass tube (Fig. 2.1) sealed at the upper end and filled with water is raised above the water surface, what effect will this have on the water potential at the upper end of the tube? When the tube has been raised so that the sealed end is 10.13 m above the surface, a pressure gauge at the top will register a pressure of −1 bar, corresponding to absolute vacuum. (We can ignore the small effect of water vapour pressure.) If the tube is raised still further the pressure, and hence the water potential Ψ_p at the top, becomes even more

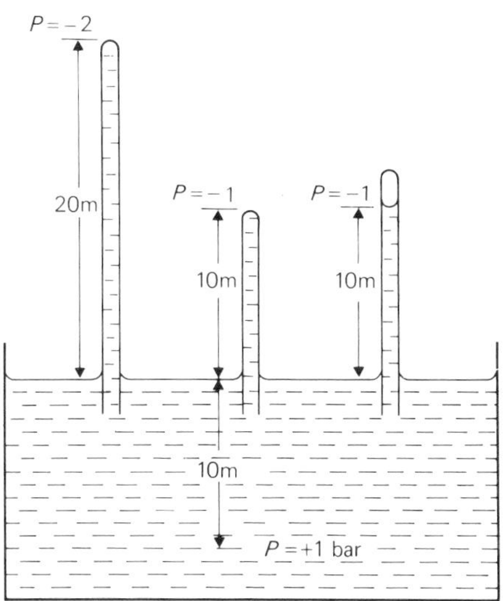

Fig. 2.1 Showing open-ended tubes filled with water. Pressure in the top of the centre tube equates with vacuum in exactly compensating an atmospheric pressure of 1 bar. Pressure in the right-hand tube is slightly greater than −1 bar owing to the presence of water vapour – it functions as a water barometer. In the left-hand tube water continuity has been maintained by adhesion and cohesion reaching −2 bar. By descending in the body of water Ψ_p becomes positive. (Note: Strictly 1 bar equals a vertical column of 10.13 m of water at 20°C.)

negative to -2, -3 bar, etc. for each increment of 10.13 m. As the tube is elevated the liquid column becomes increasingly unstable. If it cavitates, a gas phase develops which is a vacuum except that it contains a small amount of water vapour. The column of water falls back to about 10 m height, now balancing atmospheric pressure. The column then rises and falls with fluctuations in the atmospheric pressure, having become a water barometer. Such barometers are rare, being too cumbersome in comparison with similar mercury barometers; also dissolved air is liable to affect the vacuum.

This system illustrates how negative water potentials can be produced very simply by a pressure device. It also indicates how trees which may have trunks 100 m tall, like *Eucalyptus* or *Sequoia*, must maintain continuous columns of water under tension, throughout their existence despite the danger of cavitation. Sunny conditions make the trees transpire, producing even greater tensions than those caused by gravity alone. Nevertheless, sufficient water continuity persists in the trunks for conduction and hence for survival.

Berthelot's apparatus

The apparatus devised by Berthelot (1850) is an extremely simple device to illustrate *cohesion*. A thick-walled glass tube, sealed at one end and drawn out at the other, is filled with pure water. When water almost fills the tube it is sealed in a flame (a difficult operation!). If the sealed tube is now warmed the internal pressure increases, because the expansion of water exceeds that of the glass, and any remaining gas or vapour is forced into solution.

On cooling the tube, providing a gas nucleus is absent, the thermal contraction of the water produces an internal tension or negative pressure. Eventually cavitation, the sudden development of a gas phase, occurs with a sharp 'click'. A cloud of small gas bubbles is produced which rise and coalesce in the tube. This warming and cooling cycle can be repeated indefinitely. The negative pressures at incipient cavitation can be calculated on the basis of differential thermal expansion, but a more convincing indication is given by bending the tube into a coil (see Fig. 2.2*a*). Positive and negative pressures cause the coil to uncurl or curl and an attached pointer can be calibrated using positive pressures, indeed this is the principle behind standard Bourdon gauges.

Considerable negative pressures can be produced (down to -60 bar) in a Berthelot tube if filled with pure water or plant sap. The system illustrates how cohesion could be manifest in vascular tissues. Cavitation is a trigger, releasing energy stored by distortion of the walls of a liquid container. This device is used by many plants to

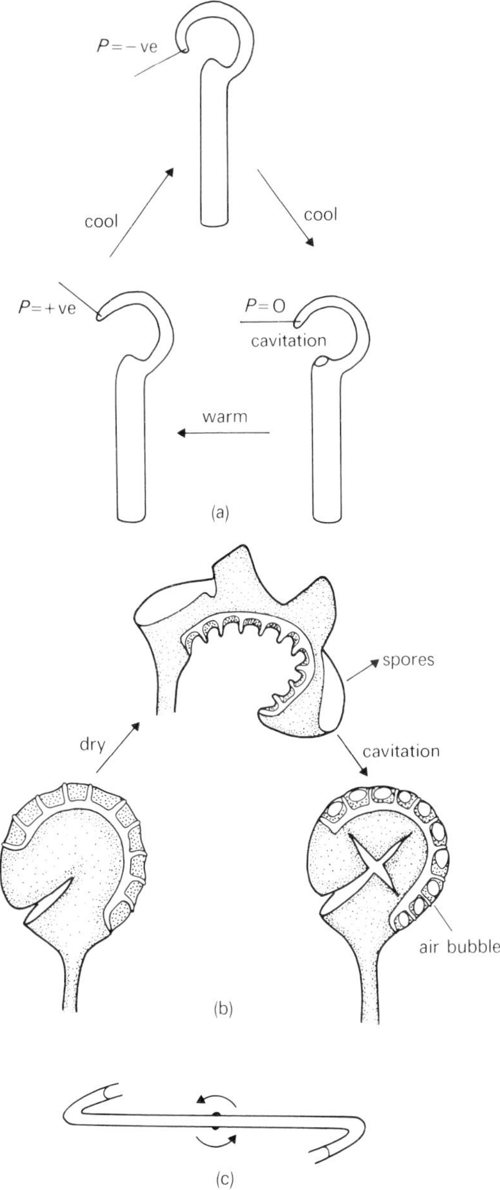

Fig 2.2 (*a*) Diagrammatic representation of the Berthelot tube fitted with a pressure-sensitive coil. When warmed the tube fills. On cooling water contracts, reducing the pressure as shown by the coil, until pressure is released by cavitation. (*b*) The principle behind a fern sporangium resembles a Berthelot tube except that water in specially thickened cells of the annulus is thrown under tension by evaporation, not cooling. Cavitation triggers release of the dried spores which are catapulted away leaving air-filled annular cells. (*c*) A water-filled Z tube with open ends retains water when spun centrifugally until cavitation disrupts the continuity.

disperse their spores. In ferns the sporangium is beautifully constructed to amplify the distortion of the walls to project spores strongly (Fig. 2.2*b*). The annulus is thrown under contraction as it dries. Eventually cohesion is disrupted by cavitation when the strained walls catapult and disperse the dried spores. At one time it was thought that gases, which are normally dissolved in plant saps, would not allow strong negative pressures to develop but this was disproved by Dixon by testing normal sap in Berthelot tubes.

Another method to induce cavitation is that of a water-filled Z tube mounted on a centrifuge head (Fig. 2.2*c*). Briggs (1950) spun a Z tube to induce cavitation centrally in the tube where the water could be subjected to pressures down to -277 bar at 10°C, calculated from the centrifugal force. Eventually cavitation disrupted the system. Curiously the limiting negative pressures fell slightly when the temperature was raised but drastically when the temperature was reduced towards 0°C, an effect yet to be fully explained. If a similar effect were to occur in plants their conduction systems could be impaired directly by the effect of low temperature.

The Askenasy apparatus

The fact that wood is made of capillary tubes suggested to early investigators that water might ascend a tree, like water in a glass capillary tube, by surface tension. In fact these tubes are far too wide to account for more than a few metres of ascent, even leaving aside the knotty problem of how a tube with a gas interface at the top could supply leaves with water. Yet the apparatus developed by Askenasy (1895) from a simpler version by Detmer showed how sap ascent might occur in a physical system which also illustrates the essential role of surface tension very clearly.

A porous pot can be imagined as a multiplicity of fine capillaries in a matrix and the apparatus (Fig. 2.3) consists of a porous pot, specially saturated with water, attached to a vertical glass tube. Both are water-filled and stood in a beaker of water. To avoid the necessity of having a very tall glass tube to generate negative tensions, Askenasy allowed the lower end of the glass tube to dip into mercury (1 bar is equivalent to a 10.13 m column of water but only 0.76 m of mercury) which also allowed him to measure the negative pressure directly. When set up, the height of the mercury column was zero, but as water evaporated from the porous pot it climbed to a height of over 1.30 m. (In Preston's (1958) experiments it reached 1.50 m, i.e. -2 bar.) Beyond this tension the system became unstable and gas bubbles formed inside the apparatus. Thut (1932) obtained results similar to

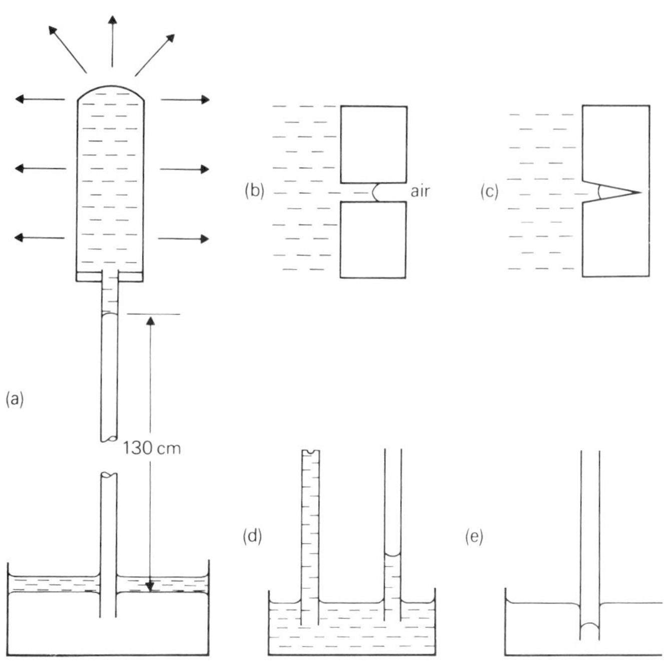

Fig. 2.3 The Askenasy apparatus having drawn up a column of mercury by evaporation of water from a porous pot. (*b*) Diagrammatic detail of a single pore in the pot showing how surface tension opposes gas pressure tending to enter the system. (*c*) A bubble in a hydrophobic crack which under negative pressure can emerge to disrupt the system by nucleation. (*d*) Two capillary tubes showing water columns held against gravity by surface tension. The tube on the left with a small pore at the top, has been filled initially with water. It now behaves like the Askenasy apparatus. (*e*) A glass capillary tube tending to repel a liquid such as mercury or oil depresses the fluid level in the tube against gravity.

those of Askenasy, except that he substituted plant shoots in place of the porous pot.

Why did air not invade the system earlier via the pores in the porous pot which must have allowed the escape of water? Apparently, providing the pores in the porous pot are sufficiently fine, gas cannot enter because of the water surface across the pore, but nevertheless water evaporates from it by diffusion. Providing the water is attracted to the pore walls, i.e. exhibiting *adhesion*, it develops a meniscus which tends to oppose the entry of air. It is important to note that air can pass freely through a dry pot or water through a wetted pot, but an interface between gas and liquid is unable to pass. Cell walls exhibit exactly the same property and, to a lesser extent, soils also. If water is repelled from a pore, as when the pore is waxed, little or no suction (ascent of mercury) develops because air invades the system almost immediately. The normal system probably breaks down when *surface tension* fails to oppose gas entry in the widest pore, which represents the weakest link in the system. Pressures which can be resisted by surface tension depend on pore diameter, thus a tension of -100 bar would require (at 20°C assuming zero contact angle) r to be no greater than:

$$P = \frac{2S}{r} \quad \text{so} \quad r = \frac{2 \times 7.3 \times 10^{-2}}{100 \times 10^{5}} = 14.6\,\text{nm} \qquad [2.2]$$

Hydrostatic pressure	$P(=h\rho g)$	$ML^{-1}T^{-2}$	Pa $(=\text{N m}^{-2} = \text{bar} \times 10^{-5})$
Surface tension fluid	S	$(ML^{-1}T^{-2})L^{-1}$	$\text{N m}^{-1}(=\text{J m}^{-2})$
Radius of tube	r	L	m

Another cause which may disrupt the system is the existence of hydrophobic cracks which contain gas able to resist dissolution. Under negative pressure a bubble can be sucked from the crack which gives a nucleus of gas so that the whole system breaks down by nucleation. Dirt particles can introduce many nucleation sites.

The apparatus is interesting in exhibiting how a physical system can cause the ascent of water beyond a pressure corresponding to vacuum (0.76 m mercury). Above this height there can be no question that water must be raised by some harnessing of atmospheric pressure and the fact that water continues to fill the system shows *cohesion*, not merely between water, but also mercury molecules. *Adhesion* between pot and water, glass and water and water and mercury is also essential for the demonstration. Disruption may be caused by a weakness at any of these points, or from the presence of dirt or grease.

The osmometer

Osmosis is the underlying principle behind an important and rather special water machine, an *osmometer*. In its original form an osmometer consisted of suitable membrane tied tightly across a glass tube and filled with solution.

In later types the membrane was deposited inside a porous pot (Fig. 2.4). When water is put in contact with the membrane it tends to enter the osmometer increasing the volume of the solution. If the volume of the osmometer is fixed, a pressure is produced. This *osmotic pressure* of a solution is defined as the pressure which must be exerted on it to just prevent any net entry of pure solvent via a

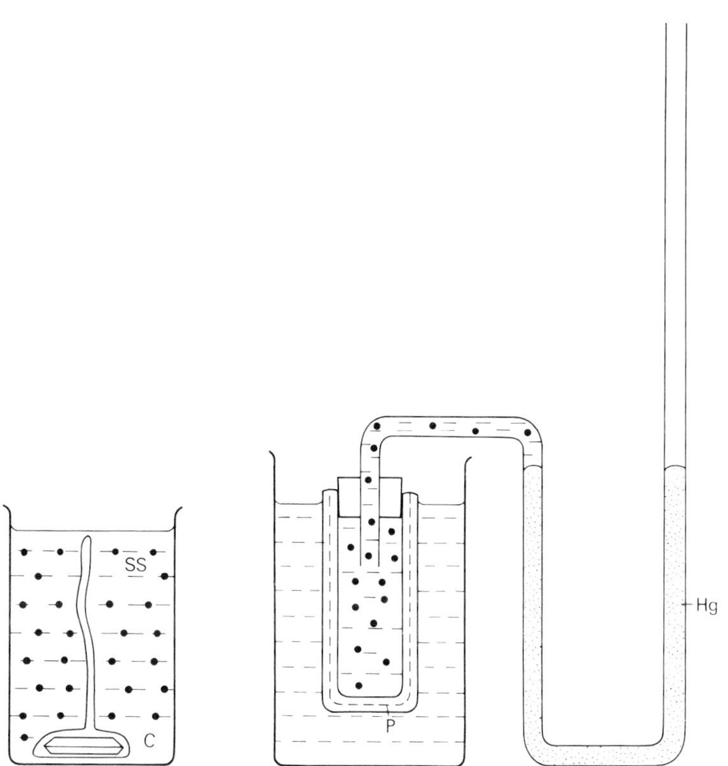

Fig. 2.4 A simple 'osmometer' (left) can be made by dropping e.g. a crystal (C) of copper sulphate in 10 per cent sodium silicate solution (SS). An osmotic membrane of copper silicate promptly forms round the crystal which after a series of bursts and automatic repairs grows a tubular outgrowth. Theoretically it could be used to compare different osmotica, in practice it is too brittle.

A conventional osmometer (right) able to resist hydrostatic pressures is made by depositing a ferrocyanide membrane within the porous walls of an unglazed pot (P) which is immersed in water. In use, the volume of solution within the osmometer would be kept constant by regulating the mercury manometer (Hg) and the apparatus would be temperature controlled.

perfect semi-permeable membrane separating solution from solvent.

It was shown by the German botanist Pfeffer that as the solute concentration increased so the tendency to attract water increased and that

$$\Psi_p = -(\Psi_s) = \text{osmotic pressure } \Pi$$

$$= \frac{1}{\text{Volume of solution } \bar{V}} \left(\text{or } \Pi = \frac{1}{K\bar{V}}\right) \qquad [2.3]$$

The similarity between this expression and the gas laws of Raoult and Van't Hoff did not go unnoticed. According to this analogy, solute molecules gave rise to the pressure by bombardment of the walls of their container as gas molecules do. The role of the solvent, water, was wrongly ignored, being equated with total vacuum. Because of this observation, osmotic phenomena have been linked with the gas laws and this explains why the gas constant R appears in the simplified equation (see Dick, 1966) for osmotic pressure thus

$$-\Psi_s = \Pi = RTm \qquad [2.4]$$

where Π is the osmotic pressure, T is the absolute temperature K, m is the molal concentration (or more strictly an activity coefficient).

The mechanism of osmosis

The incorporation of the gas laws into osmotic phenomena has on balance introduced almost as many complications as it has resolved. To begin with the driving force in an osmometer is derived not from the solute molecules (corresponding with gas molecules in the analogy) but from the solvent molecules, i.e. water. For an osmometer to work efficiently the pores in the membrane must be semi-permeable, i.e. allow the passage of water but not the solute (called the *osmoticum*). Water flows across the membrane to produce an osmotic pressure, flowing from a high water potential to a low water potential which can be considered mostly simply in terms of water concentration as explained in Chapter 1 (see Fig. 2.4).

We can understand the mechanism better by taking an extreme example. Imagine an osmometer fitted with a pressure gauge filled with dry, finely-ground sucrose powder. Despite the presence of the solute molecules, no pressure is recorded until the membrane surface is immersed in water. A pressure develops because some water molecules enter the membrane forming hydrogen bonds with the sucrose molecules. Flow continues because the remaining water molecules in the osmometer are still more free to form hydrogen bonds than the molecules in pure water so generating a localised suction adjacent to the osmotic membrane. If the volume is fixed the osmometer registers a hydrostatic pressure which builds up until at equilibrium net water influx from osmosis is balanced by the hydrostatic expulsion of water. At this point the hydrostatic pressure is called the *osmotic pressure.*

The energy of the bonds tending to bind sucrose and water may be visualised by imagining the heat energy required to evaporate or freeze water from a solution to restore its capacity to produce an osmotic pressure of the initial value.

According to the explanation above, therefore, osmosis is the harnessing of a water potential gradient to produce either a flow or a hydrostatic pressure in an osmometer. If we assume there are pores in the membrane they must be filled with pure water. A mass flow through the pores must be induced by a local reduction in pressure at the mouths of the pores of the membrane surrounded by solution.

Regarding terminology, opinion varies. Some would say that a solution in a beaker has an osmotic pressure of a given value. There is no measurable hydrostatic pressure; they mean that if the solution is enclosed in an osmometer against water a hydrostatic pressure *could* be produced. It seems better to describe such a solution as having a solute or *osmotic potential* Ψ_s which is *identical* numerically and in sign with the lowering of the water potential by solute addition. Thus a 1 molal solution of sucrose has an osmotic potential Ψ_s of -26.3 bar which equals the water potential of -26.3 bar. If placed in an osmometer against water, an osmotic pressure Ψ_p (or $-\Psi_s$) of 26.3 bar is produced hydrostatically (note that we do not say bars because the 's' would indicate seconds) at equilibrium.

Thus $\Psi_p = -\Psi_s$ because at equilibrium $\Psi = \Psi_p + \Psi_s$ and Ψ_s is always negative.

Large particles suspended in water have an insignificant osmotic potential, but smaller *colloidal* particles, even of metals like gold, have a measurable osmotic pressure. Large dissolved molecules such as protein (e.g. gelatine) are much less capable of generating osmotic pressures than an equivalent *volume* of smaller molecules such as an electrolyte or glycerol. Among true solutions, such as sugar, concentrations of equal molality have approximately the same osmotic potential around -22.4 bar at 20°C of a 1 molal solution because the molecules are not ionised. However in ionised solutions, such as sodium chloride, the number of particles and hence the extent of hydrogen bonding is almost doubled (it is reduced slightly by electrostatic attraction between the ions, see Dick, 1966), so that a molal solution has an osmotic potential of -44.5 bar at 20°C. It is a curious fact that osmotic pressures develop so nearly in proportion to the number of solute molecules or ions, irrespective of their species. Possibly this is because they all attract similar shells of water molecules. In general it has been customary to regard an osmoticum as 'ideal' if it obeys the gas law prediction, hence a 1 molal solution ideally produces an osmotic pressure of 22.7 bar equalling 1 osmole, a term frequently used in animal physiology and medicine. Solutions rarely conform with this ideal state, so that an osmotic coefficient ϕ is used in addition to correct the gas law equation thus:

$$\Psi_p = -\Psi_s = \Pi = \frac{zRTM\,m\,\phi}{\bar{V}} \qquad [2.5]$$

Osmotic pressure	$\Psi_p = -(\Psi_s) = \Pi$	$ML^{-1}T^{-1}$	Nm^{-2} (=Pa)(=bar $\times 10^{-5}$)
Valency	z	–	–
Gas constant	R	$ML^2T^{-2}\,mol^{-1}\,K^{-1}$	$J\,mol^{-1}\,K^{-1}$
Absolute temperature	T	K	K
Molecular wt solvent	M	$M\,mol^{-1}$	$kg\,mol^{-1}$
Molality	m	$mol\,M^{-1}$	$mol\,kg^{-1}$
Osmotic coefficient	ϕ	–	–
Partial molal volume	$\bar{V}$	$L^3\,mol^{-1}$	$m^3\,mol^{-1}$

Osmotic coefficients of electrolytes are usually slightly less than unity for the reason given above. Proteins may have osmotic coefficients as high as 2.45 (Dick, 1966). One reason for this is that some water is bound internally within the molecules so ceasing to behave as an ordinary solvent. Several osmotic coefficients are tabulated in Appendix 10.

Finally we may observe that if in the apparatus in Fig. 2.4 we had placed water *inside* the osmometer and the solution *outside*, the osmometer would register a negative hydrostatic pressure or tension. At equilibrium this is a direct measurement of the difference between the internal pressures of pure water and the solution. The reason we do not set up osmometers in this way is that human technology has not mastered the measurement of considerable tensions without disruption (cavitation). Nevertheless plants function in this way most of the time!

Porous systems: the reflection coefficient σ

It is commonly assumed that a semi-permeable membrane is essential for osmotic behaviour, but this is not the case, porous systems can also behave in this way. During experiments in which leaf discs were transferred from osmotic liquids to water it was observed that droplets often exuded from the surfaces of the discs. Model discs made of thick card, waxed on both sides, were tested similarly. First they were floated on strong salt solution. No exudation was detected. On transferring the discs to water, however, droplets of exudate burst through the waxed surfaces in exactly the same manner as in leaf discs (Fig. 2.5).

Apparently cell walls or wet cellulosic material like paper are sufficiently fine in texture to behave like a leaky semi-permeable membrane. Ions slowly penetrate the discs floated on salt solution, but after transfer, water is able to penetrate the discs more rapidly than the ions can escape, building up in volume and causing osmotic

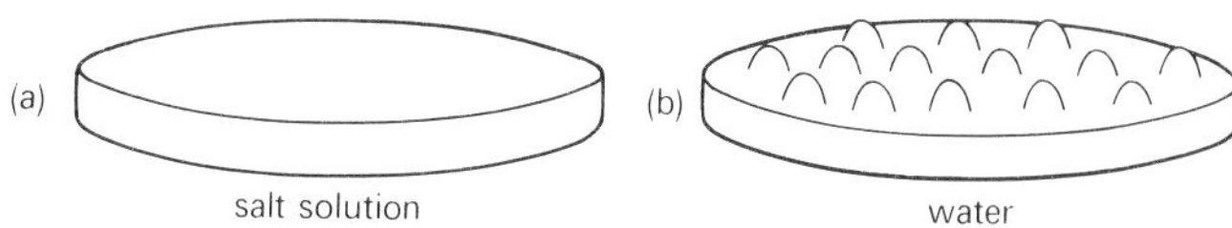

Fig. 2.5 Exudation of droplets from a porous 'osmotic resistor' of card floated initially on saline solution (*a*) then water (*b*).

exudation. Such an effect is of course temporary. In principle any porous matrix can behave in a similar manner (cell walls, soil materials, gels, etc.). When gels absorb water considerable pressure can be generated, as in the experiment shown in Fig. 2.6. Dried peas, packed into bottles known to burst at a pressure of 19 bar, burst them by absorbing water. In a similar way dry wooden wedges have been used to split rocks, such as granite, since antiquity. Plants often use this mechanism and this allows plant roots to fracture concrete and asphalt.

Fig. 2.6 'Bottle bursting' as dried peas, packed into glass bottles with perforated stoppers, take up water. One burst, the other did not, probably reflecting differences in packing. Plant root tips penetrate soil and fracture asphalt rocks and concrete in a similar manner.

Porous systems, and many natural membranes, do not behave like perfect semi-permeable membranes but to a certain extent they 'leak'. The measure of this leakiness of a differentially porous system (membrane) for a particular solute is the reflection coefficient σ, which ranges from unity, for a perfect membrane transmitting water only, to zero where the membrane is so leaky that no osmotic behaviour is demonstrable. A formula expressing the osmotic pressure developed in such a leaky system is

$$\Pi = \frac{nRTM\,\phi m\sigma}{\bar{V}} \qquad \text{where } \sigma = \frac{\Psi_p \text{ developed}}{-\Psi_s, \text{ driving it}} \qquad [2.6]$$

and where the parameters are the same as described above.

Perpetuum mobile machines?

For many years an intriguing puzzle in biology has been to decide the extent to which plants and animals move water without recourse to osmotic phenomena. The observation that plants exuded almost pure water from leaves during guttation (see Ch. 4) suggested that this could not be an osmotic phenomenon. Similarly when Van Overbeek (1942) noticed a disparity between the osmotic composition of exudate from decapitated root systems and the osmotic potential of an external medium required to prevent the exudation, he suggested that this might be due to 'active' water transport.

A simple explanation for many such phenomena is that an osmotic pressure is indeed generated adjacent to semi-permeable membranes. This pressure induces solution flow in bulk. Finally the osmotic solutes are reabsorbed and cease to behave osmotically before the liquid is released. A model of this kind which absorbs pure water at one end and exudes it at the other can be constructed simply (Fig. 2.7) from an osmometer containing salt solution with cation and anion exchange columns in series. In whole plants solutes are absorbed by living cells but water cannot follow the solutes on account of the turgor pressure which increases. In the model, energy is provided to concentrate the solute and to recharge the exchange columns. A similar expenditure of energy is also required to power such flow in plants. Flow is not, of course, perpetual and declines towards zero as Ψ_s of the osmoticum rises towards zero as it is diluted by water.

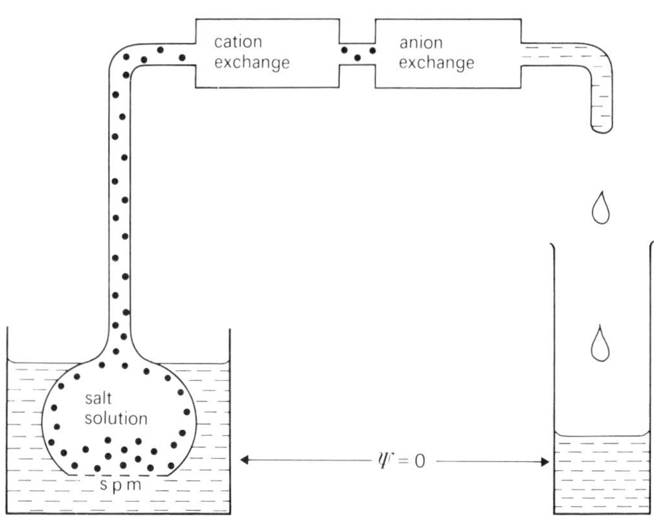

Fig. 2.7 A *perpetuum mobile* machine in which water is raised against gravity despite the absence of a water potential gradient overall. Of course flow slows down and finally stops as the water potential inside the osmometer is raised as a result of the water influx.

The occurrence of active water transport

Active water movement can be defined as the net movement of liquid water (pure or in solution) against a Ψ gradient and *driven directly* by metabolic activity. The intervention of osmosis, as in the model above, would not qualify as active movement. Some of the most promising and well documented examples are from the animal world. For example a gall bladder preparation can be made which pumps water between two media even though their osmotic potentials are initially identical. Apparently in this case (as in the case of guttation also), though an osmotic mechanism is not obviously involved, it *is* the basic mechanism driving flow. Osmotic flow and reabsorption take place within the walls of the gall bladder which thereby seems to transport water alone.

Another intriguing system is the insectivorous bladder of the *Utricularia* plant which is primed by expulsion of water. Sydenham and Findlay (1975) have shown, however, that active ion transport sets up an osmotic gradient which moves the water passively; the quantitative aspects are not yet resolved. It seems, therefore, that if water molecules are moved (by utilising the hydrogen bond as a 'handle'), the mechanism is indirect and some form of osmotic or electrical activity is always involved.

The only other method for moving water is in bulk. Thus when a giraffe stoops to drink, water is raised against gravity to its stomach. (Even man, when upside down, can elevate water by this peristaltic mechanism.) It is very probable that such active water movement does occur in this form in many systems including plant and animal cells driven by peristalsis or pinocytosis. Such a mechanism has still to be studied in detail in plants, but its capacity to develop a large pressure gradient is doubtful. The same cannot be said, however, of certain insects which have long been known to be capable of gaining in weight at the expense of atmospheric humidity. The champion water-extractor is *Thermobia*, the firebrat, which is capable of extracting water from air to around 50 per cent relative humidity, the critical limiting humidity. ($\Psi = -1{,}000$ bar!) Since Ψ_s of the body fluids is only around -10 bar this represents a truly remarkable achievement! The site of uptake was shown by Noble-Nesbitt (1975) to be the rectum, which functions as a water extractor in most insects. In this case the insect pumps moist air inwards and ejects drier air by rhythmic body movements. Plants have not been shown to possess the capacity to actively extract useful quantities of atmospheric water. Atmospheric water can be absorbed by *Tillandsia usneoides* (Spanish moss), a rootless epiphyte which lives on trees and relies on atmospheric water and nutrients. However, this has proved to be a function of dead rather than living tissues (De Santo *et al.*, 1976). But how useful it would be if desert xerophytes could use solar energy to extract atmospheric water for their growth and survival!

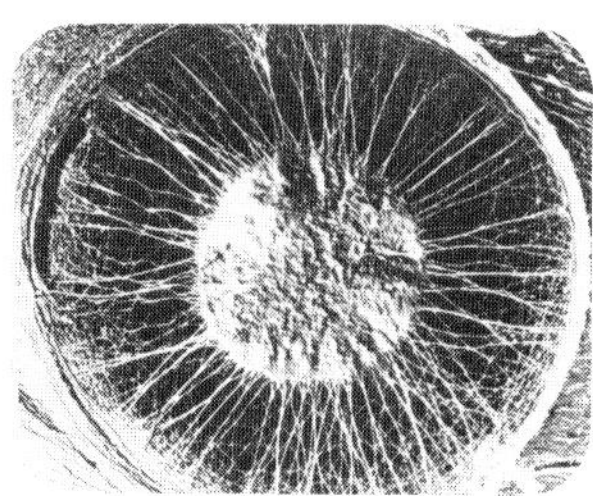

Chapter 3

Water relations of cells

Unlike animal cells, plant cells have a tough but porous outer casing, the cell wall, containing a protoplast and often a large prominent vacuole. The living material of such a cell, the cytoplasm, contains practically all the organelles and lies close to the cell wall. In immature cells the nucleus is often central. As vacuolation proceeds it may remain suspended centrally on connecting cytoplasmic strands, but more usually it too lodges near the cell wall. The vacuole may occupy over 90 per cent of a parenchyma cell, but a much smaller fraction, around 40 per cent, of mesophyll cells packed with chloroplasts.

Two membrane systems can be discerned with the optical microscope. The outermost membrane, the plasmalemma bounding the protoplast, is only 5–10 nm thick but its position can be discerned easily when a cell has been plasmolysed. An inner membrane, the tonoplast, bounds the vacuole. The plasmalemma seems to be the most important membrane in regulating water relations, but the tonoplast, in separating living cytoplasm from the fluids in the vacuole (which include excretory products) may well serve a protective function. Cells are interconnected by plasmodesmata, despite intervening the cell walls which often seem to act as barriers, so that the cytoplasm of a plant can be considered as a continuum, called the *symplast* by Arisz. Numerous organelles, e.g. mitochrondria and plastids, are also bounded by semi-permeable membranes which cause them to respond osmotically. Furthermore the endoplasmic reticulum is an organelle consisting almost entirely of a complex network of fine tubular membranes (Fig. 3.1).

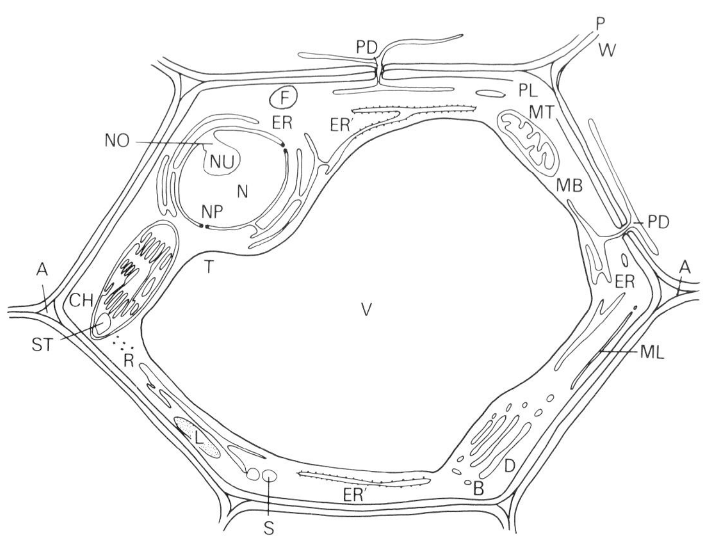

Fig. 3.1 Diagram to illustrate the ultrastructural organelles in a typical vacuolated plant cell. Their probable function is indicated. Airspace A (gas exchange), Golgi body B (protein vesicle), chloroplast CH (photosynthetic assimilation, dictyosome D (Golgi apparatus, export packaging), smooth ER and rough ER′ endoplasmic reticulum (lipid synthesis, sugar transport?), fat body F (storage product), lysosome L (enzyme pack), microbody MB, microtubule ML, mitochondrion MT (ATP synthesis by respiration), nucleus N (control and replication), nuclear organiser NO, nuclear pore NP, nucellus NU, primary wall P (rapid water flow), plasmodesm PD (intercellular transport), plasmalemma PL (import export control water and solutes), polyribosome R (protein synthesis), spherosome S (ER production?) starch grain ST (food reserve), tonoplast T (vacuolar control), vacuole (excretion and support), secondary wall W (water flow and support).

When a cell is taken from a multicellular plant it must be borne in mind it is a single, often specialised, member of an integrated team of cells, called a tissue. Most cells become damaged to some extent during isolation before we can examine them in detail. Nevertheless by studying the behaviour of individual cells, and ultimately of individual organelles a picture of the integrated system can be gradually pieced together. This must be tested in turn in whole-plant physiological experiments before we can be sure that our reconstruction is correct.

The single cell

The following description will apply particularly to a parenchyma cell (literally a 'packing cell') with which the higher plant world abounds. It is generally regarded as 'typical' of other cells, but since it has thinner walls, larger vacuoles and a low percentage volume of organelles, this assumption can be challenged.

Cell wall

The primary cell wall consists of about 5 per cent cellulosic crystallites, 2 per cent protein, 33 per cent pectic and allied substances and about 60 per cent water contained, at least in part, in microcapillaries 1–5 nm diameter. On this wall are laid additional layers comprising the secondary walls, which may skirt, at intervals, areas of wall to leave pits. The additional layers normally run across each other to give a strong laminated structure able to withstand considerable tensions or internal pressure within the cell. Gels, consisting of pectins and allied substances, fill the matrix of cell wall which is bound by glucans and related compounds. Cytoplasmic structures occur in the walls especially in plasmodesmata (some are claimed to occur on the surface of a plant called ectodesmata). Osmotically active fluids, stains, ionic solutions, etc. have ready access to the plasmalemma membrane within the relatively porous walls.

Membrane structure

Early studies on all membranes showed that hydrocarbons and other fat soluble (lipophilic) substances penetrate cells with surprising ease. Also when cells are treated with reagents such as chloroform or ether the membranes become leaky and sap runs out (slices of beetroot are often used to demonstrate this in classes, because the escaping sap is coloured with betacyanins and therefore readily measurable). These observations indicate that membranes are part lipid. This view has been confirmed and extended and phospholipids and sterols (e.g. cholesterol) are known to be a major constituent of membranes. Protein seems to be the other main constituent combining somehow with lipids to form a layered structure. From electron microscopy at magnifications around 200,000 times, a 'unit membrane' (named by Robertson in 1964) is visible as a double layer 6–8 nm thick: indeed most membranes seem to be about 7.5 nm across. The way the protein molecules intermesh with lipids has been a puzzle. One suggestion made by Daveson and Danielli is that the lipids form the filling of a sandwich with proteins on each side. A more modern concept is that the lipids may form a hexagonal matrix studded with protein molecules (Fig. 3.7). In recent years it has become increasingly clear that membrane structure and permeability can be strongly influenced by both pressure and electrical gradients. It seems possible that hormones may influence membrane permeability by modifying either pressure or electrical potentials.

Cytoplasm

The cytoplasm with the nucleus is regarded as the living substance of the cell. Though largely bounded by tonoplast and plasmalemma it is really more diffuse than it appears with tentacle-like processes passing through cell walls (*plasmo-* and *ectodesmata*) and strands traversing the vacuole. Its matrix may be aggregated to form a relatively immobile *gel*, or a freely fluid *sol* state. Normally cytoplasm consists of a shifting equilibrium between these states and when viewed under an optical microscope streaming can often be seen in a constantly shifting circulatory pattern superimposed on the purely physical agitation of cellular particles called Brownian movement. Cytoplasmic particles include chloroplasts or non-pigmented plastids, mitochondria, Golgi apparatus, ribosomes, dictyosomes, etc. (Fig. 3.1) and can only be seen in detail with electron microscopy. In recent decades, and particularly since the advent of the electron microscope, there has been a shift in concept from the view that cytoplasm is a rather simple gel, to the realisation that it is a complex and highly organised matrix nevertheless capable of contractility, streaming and many complex biochemical reactions. Energy is provided by respiration in mitochondria which powers biochemical syntheses, solute transport and streaming movement.

Vacuole

Immature cells lack obvious vacuoles (meristematic cells have microvacuoles) but during expansion usually several vacuoles become recognisable which gradually increase and coalesce so filling the centre of the cell. Numerous solutes are present in the vacuole including inorganic ions and complex organic molecules. These solutes are responsible for the osmotic potentials of cells, which range from about 5–20 bar, and also the pH of cell sap which is usually acidic, around 4–6. Many cells have coloured vacuolar sap. Anthocyanins are common in many epidermal cells and storage tissues also, e.g. the beetroot. Other vacuoles may contain tannins, terpenes, latex, crystals of calcium oxalate or carbonate and some (e.g. *Desmarestia*) even contain sulphuric acid. The presence of these substances suggests that vacuoles serve an excretory function.

Cell expansion and turgor is governed in part by the extent to which the water potential is lowered by the presence of solutes which thus regulate the internal hydrostatic pressure. We now know that this turgor pressure is controlled by several different mechanisms. Thus cells in the pulvinus of *Mimosa pudica* (the 'sensitive plant') act as a hinge supporting the leaf. In response to mechanical irritation the membranes of the cells become permeable allowing the vacuolar contents to escape into the airspaces. This loss of vacuolar sap causes cellular collapse which allows the leaf to fall. In a similar way *Rhododendron* leaves droop when petiolar cells adapt to frosty conditions. Ions may be moved at the expense of metabolic energy to provide rapid opening and closure of stomata (see Ch. 7). Disaccharide sugars such as sucrose can be cleaved enzymically into hexoses nearly doubling their osmotic potential. It seems that vacuoles may directly influence fluid movement by contraction. Many algal and animal cells possess contractile vacuoles which seem to serve as 'bilge pumps', expelling a surfeit of osmotic water at regular intervals.

Cell water relations

In this section we will examine the more extreme effects of the removal of water from cells as a prelude to studying cells with more normal water deficits.

Plasmolysis

In the laboratory the water potential around a cell can be regulated artificially by accurately using solutions of appropriate Ψ_s. When Ψ_s of the bathing solution is reduced below the hydrostatic pressure at full turgor the cell becomes *plasmolysed.* During the plasmolysis the protoplast of a cell shrinks leaving the wall behind. Plasmodesmata, the cytoplasmic threads connecting cells through cell walls, are pulled, torn and snapped; new membranes usually form to seal these ruptures. Depending on plasmodesmatal connections the protoplast may become concave or convex in shape. Plasmolysis resembles wilting (see below) except that the bathing solution penetrates the cell wall, displacing the protoplast. Unlike wilting, plasmolysis occurs quite rarely in nature. The protoplast shrinks progressively but the volume enclosed by the cell wall ceases to decrease when plasmolysis is reached, contrasting with the wilted condition (see inset Fig. 3.3).

It is important to select an osmotic solute carefully. Plasmolysis can only be produced if water permeates an intact cell more rapidly than the solute. Some solutes like methanol or ethanol, penetrate a protoplast rapidly and simultaneously coagulate (fix) the proteins. Others, such as methyl urea, can penetrate so rapidly that plasmolysis is reversed in several minutes even without changing the bathing solution. Sucrose is frequently chosen because it has relatively slow penetration rate compared with water. The sugar alcohol mannitol is also a favourite osmoticum based on the fact that it is usually much less readily metabolised than sucrose which also appears to reduce its permeability. Some plants actively metabolise mannitol, however. The permeability of a particular cell to take up a solute is measured as the reflection coefficient σ (see Ch. 2). Clearly for a particular solute σ can be measured by observing the rate of recovery of a cell from plasmolysis when kept in the same bathing solution. In general the lower their relative molecular masses the more rapidly do solutes penetrate living cells. The relative molecular mass of solute gives an indication of its molecular dimensions. Lipid solubility of a solute has an important effect, however, apparently indicating that some solutes and lipid solvents penetrate cells by partitioning within the lipid (fatty) component of the cell membranes. Similarly if cells are treated in a lipid solvent, for example, chloroform vapour, they eventually lose their semi-permeability altogether and the vacuolar contents escape freely. Probably the integrity of the membranes has been destroyed because the lipid component has dissolved.

Plasmolytic methods require only a light microscope and appropriate solutions as equipment and consequently have provided a convenient method for quantitative studies on all water relationships.

Wilting

Cells exposed to air tend to lose water by evaporation very readily. Even if the air is very humid it still has a very low water potential. At 20°C air in the range of 100 to 90 per cent relative humidity corresponds with an osmotic solution, with a water potential of 0 to -26 bar. Air is difficult to maintain in this high humidity range because even small temperature gradients tend to lower both humidity and water potential. For this reason, it is far more convenient to regulate cell water with osmotic solutions, rather than moist air even though the latter is more natural.

A cell losing water in air crumples as a whole and wilts. Wilting is quite different from plasmolysis in which osmotica freely penetrate the cell wall causing it to separate from the shrunken protoplast (see inset Fig. 3.3). In air the moist porous cell walls resist the entry of air on account of the surface tension of water filling the minute pores. Consequently the whole cell usually crumples and the wall and protoplast remain in contact. A few cells, such as *Sordaria* spores (see below), have rigid walls and in such cases negative turgor pressures develop instead of wilting.

Most plant cells are only slightly deficient in water and not so deficient as the extreme examples described above. Cellular physiology is greatly influenced by even small changes in water potential, so how can these changes be measured?

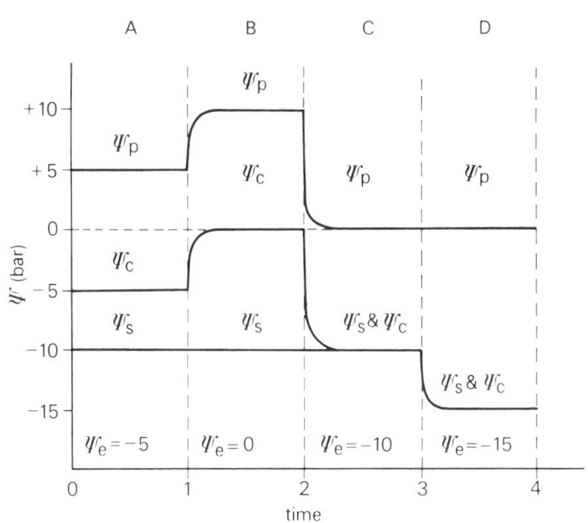

Fig. 3.2 Hypothetical experiment conducted on an inelastic parenchyma cell (A) intact in a living plant with environmental water potential Ψ_e of -5 bar corresponding with a xylem sap tension of -5 bar. Ψ_e counteracts cellular Ψ_s, but the water potential still maintains a turgor pressure Ψ_p of 5 bar (B). On transferring to water $\Psi_e \rightarrow 0$. Ψ_{cell} also$\rightarrow 0$ by adjusting Ψ_p to exactly compensate Ψ_s. (C) If the cell is immersed in an osmotic fluid such that $\Psi_e = -10$ bar incipient plasmolysis occurs; $\Psi_p \rightarrow 0$ and $\Psi_{cell} = \Psi_s$. (D) When more solutes are added to the bathing medium, Ψ_e becomes -15 bar, Ψ_p is zero because the cell walls separate from the protoplast. Now $\Psi_{cell} = \Psi_s$ and the cell is plasmolysed. Note that Ψ_p changes when the cell is turgid but when turgor is lost only Ψ_s changes.

The measurement of water potentials of cells or tissues

In considering the interplay of cellular water potential components (Ψ_p and Ψ_s) which produce the resultant water potential of a cell, Ψ_{cell}, it is simplest to imagine the cell is inelastic (see Fig. 3.2). Most plant cells are quite elastic however, and to obtain a clear picture of this more complex interaction it is desirable to construct the curves in Fig. 3.3 from measurements and observations, as described below.

Osmotica bathing single cells

Imagine a typical parenchymatous cell in which, after excision, Ψ_{cell} is -1.59 bar. It is observed on a microscope stage while immersed in an osmotic bathing medium which can be altered. The external solute concentration is now raised until the cell shrinks and reaches the

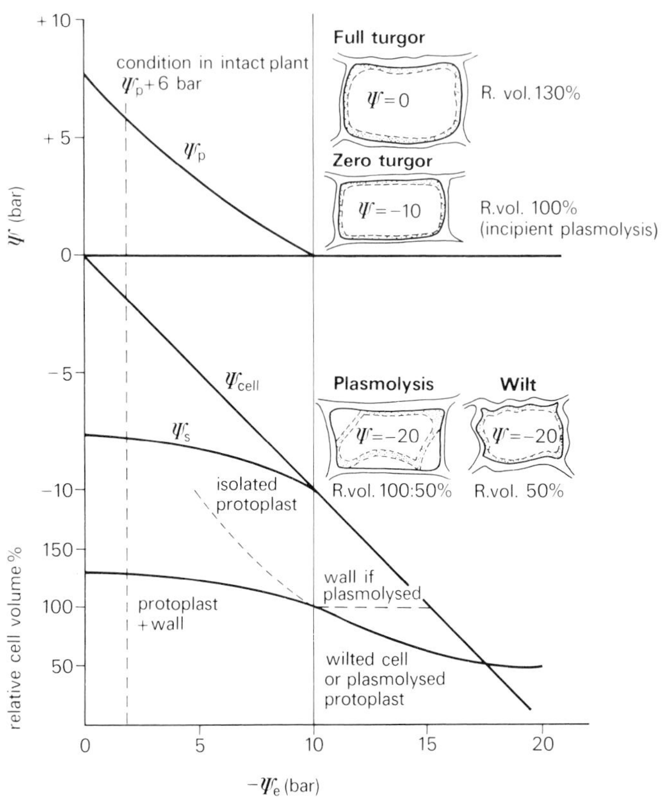

Fig. 3.3 Diagram showing how the water potential Ψ_{cell} of a typical (elastic) plant cell is the resultant of pressure Ψ_p and osmotic Ψ_s components. The volume changes of the turgid cell are subject to the cell wall elasticity. The interplay between wall extensibility and solute concentration governs the hydraulic capacity of the cell.

point of *incipient plasmolysis*, i.e. when the protoplast is just about to shrink inwards from the cell wall. At this point the turgor pressure Ψ_p, initially +6.16 bar, has fallen to zero and Ψ_s, the water potential of the vacuolar sap balances the external water potential Ψ_e which is −10 bar. The volume of the cell decreases during the process, the extent depending on the cell wall elasticity and the water potential of the cell. For the following calculations the relative volume (RV%) of the cell at incipient plasmolysis will be taken as 100 and the starting condition as 129 per cent. The volume of the cell wall will be ignored and we will also assume that the non-osmotic volume, representing the volume of starch grains, etc. is insignificant.

If Ψ_e is then adjusted to -7.5 bar the cell takes up water and is observed to expand in volume again to 110 per cent. Ψ_s of the vacuolar sap increases (becomes less negative than -10 bar) in water potential on account of the volume change and the new Ψ_s can be calculated:

$$\Psi_s = \Psi_{s(i.p.)} \times \frac{100}{\text{RV per cent}} = -9.09 \text{ bar}, \qquad [3.1]$$

Ψ_s is the new osmotic potential, $\Psi_{s(i.p.)}$ is the osmotic potential at incipient plasmolysis and RV per cent is the per cent of the cell volume at incipient plasmolysis.

The discrepancy between the internal and external osmotic potentials can be determined from the formula adapted from Eq. [1.1]:

$$\Psi_e = \Psi_{cell} = \Psi_p + \Psi_s$$

Thus $\Psi_p = \Psi_{cell} - \Psi_s = 1.59$ bar *which is the hydrostatic pressure (turgor pressure) generated in the cell partially counteracting the internal osmotic potential. This and subsequent calculations are shown graphically in Fig. 3.3.*

Similarly for $\Psi_{cell} = -5.0$ RV = 120 per cent $\Psi_s = -8.33$, so $\Psi_p = 3.33$
Similarly for $\Psi_{cell} = -2.5$ RV = 128 per cent $\Psi_s = -7.81$, so $\Psi_p = 5.31$
Similarly for $\Psi_{cell} = 0.0$ RV = 130 per cent $\Psi_p = -7.69$, so $\Psi_p = 7.69$

Of course when $\Psi_e = 0$ *this means that the cell is immersed in pure water when* Ψ_s *is exactly compensated by* Ψ_p *at equilibrium and the cell is at maximal or full turgor, i.e. the highest turgor pressure it can generate. Note that because the volume has changed, increasing to* 130 per cent RV, *the* Ψ_s *is significantly less negative at full turgor than at incipient plasmolysis.*

What, we may inquire, would have happened if the concentration of external solute had been increased so that the cell lost more water beyond the incipient plasmolysis stage? If Ψ_e *became* -15 bar *the cell would lose water until its water potential was also* -15 bar. *The wall appears unchanged because it is permeable to the osmoticum but the protoplast shrinks osmotically until* Ψ_s *becomes* -15 bar, *and the cell is plasmolysed. The protoplast volume*

$$\text{RV per cent} = \frac{\Psi_s}{\Psi_e} \times \frac{100}{1} = 66.7 \text{ per cent} \qquad [3.2]$$

Similarly if Ψ_e is made -20 bar, protoplast RV per cent becomes 50 per cent (and $\Psi_s = -20$)
Similarly if Ψ_e is made -40 bar, protoplast RV per cent becomes 25 per cent (and $\Psi_s = -40$)

From the first calculations it can be seen that, had not the cell wall intervened, when Ψ_e *was* -5 bar *the cell* RV per cent *would have become* 200 per cent *and not a mere* 120 per cent! (*It is for this reason that when protoplasts are isolated enzymatically or by cutting the walls of plasmolysed cells to liberate them, they must be kept in suitable osmotica to stop them bursting.*) *From this type of calculation we can see that the relationship* RV per cent–Ψ_e *is sigmoid composed of two opposing hyperbolic curves, one from cell wall extensibility; the other from osmotic solute-concentration effects on the protoplast.*

If, instead of studying a single cell microscopically, pieces of tissue are bathed in osmotic solutions very similar results are obtained. The sigmoid curve for Ψ is characteristic and may be derived from weight, volume or even length measurements. It is interesting to note that providing the Ψ_e at incipient plasmolysis and RV per cent changes are known, *all other curves in Fig. 3.3 may be calculated.* It is also worth noting, however, that the penetration rate of solutions into tissues, or solutes into cells, can produce errors. Nor is it unknown for the osmotic potential of vacuolar sap to be decreased metabolically in response to a low water potential.

Though in the laboratory we frequently use osmotic solutions to regulate the water potential of cells and tissues, this is decidedly artificial and largely a matter of convenience. Under natural conditions Ψ_e would correspond with moist, but not completely saturated, air. Under such conditions cells do *not* become plasmolysed, they *wilt.*

Isopiestic techniques

In the preceding section we saw how it was possible to determine the water potential of a cell by studying its response to solutions of differing water potential. Essentially the same technique can be used to determine the water potential of pieces of tissue which are regarded as a sufficiently large aggregation of single cells to remove the need for microscopic techniques. These are isopiestic techniques (Greek: same pressure) which are based on determining the point at which there is neither water influx nor exodus from a piece of tissue by bathing a piece of tissue in an appropriate solution. In practice several similar pieces of tissue are bathed in a range of solutions of differing water potential. The change in weight (or volume or length) of the tissue slices is measured after a period of time (10–30 minutes). By plotting such changes against water potential a point of zero change, the isopiestic point, is found giving the water potential of the tissue which exactly balances the external water potentials of the bathing solution. Alternatively the concentration of the bathing solution may be measured conveniently using a refractometer or the

Shardakov method (see Ch. 1) to see if it has become more or less dilute and the isopiestic point determined as above. Errors can arise from the uptake of solutes from the bathing medium by cells, so modifying Ψ_s. Sucrose solutions are recommended with minimal exposure to bathing media for the results to be clear cut. The escaping sap can interfere with accuracy if many cells are wounded when the tissue is sliced.

Pressure bomb

In recent years pressure equipment has been used to measure and alter tissue water potentials even more reliably than osmotica. This important tool is the pressure bomb. The pressure bomb was devised in its modern form by Scholander *et al.* (1965) as a means to study sap tensions in xylem (see Ch. 5). Providing the cells adjoining the xylem are at equilibrium and the xylem sap has insignificant Ψ_s, as is usually the case, the water potential of tissues can be measured directly by applying a balancing pressure to cause incipient exudation as described below. The pressure bomb has also proved invaluable for studying cell water-relations of whole tissues at different water potentials. A fully turgid leaf is prevented from losing water by enclosing it in a plastic bag. It is then sealed in a pressure bomb so that the leaf stalk projects. Air round the leaf cells is compressed gradually, so that the cells are squeezed. Sap exudes from the projecting stump. Exudate from the leaf stalk is collected until it ceases to exude and then weighed. The compressed air-pressure P of $+2$ bar is numerically equal to the newly imposed xylem sap tension ($\Psi_p = -2$ bar when the leaf is at atmospheric pressure) for which a volume V_2 of water has been collected. When the flow ceases the pressure is raised to 4, 6, 8 bar, etc. and V_4, V_6, V_8, etc. collected. The water potential of the xylem equals the water potential Ψ of associated cells at equilibrium.

For the cell protoplasts $PV = \text{constant}$, so for protoplasts a plot of $1/P$ against V is linear.

At low bomb pressures, however, the cell wall causes interference with protoplast expansion. The point at which this occurs can be detected, because the plot departs from linearity. This point corresponds to incipient plasmolysis as described above. If the linear portion is extrapolated to zero potential, where $1/\bar{V} = 0$, this gives the osmotic potential Ψ_s of the vacuolar sap at maximum turgor. This determination depends on the exuding sap being almost pure water as is usually the case; if not, Ψ_s must be increased appropriately.

It will be noted that we have been able to use a whole organ to determine the mean Ψ_s of its individual cells. This is only possible because the xylem conduction system has a relatively low resistance to flow, effectively uniting the majority of the cell population (but see Ch. 8).

Solute potential component Ψ_s

Techniques described above have allowed Ψ_s to be measured both microscopically (or using a pressure bomb) by controlling the water potential by the application of osmotica (or hydrostatic pressure). Ψ_s in cells at normal turgor pressures must be calculated from volume change measurements (using Eq. [3.2]) and Ψ_s at the point of incipient plasmolysis or incipient wilting.

The same principle can be applied to tissues which are sufficiently transparent to be examined microscopically such as cut sections or epidermal strips. The tissues are placed in a range of different osmotica. Since a tissue is a population of cells not all cells are affected equally by a given osmoticum. Conventionally a graph is drawn of the percentage of cells plasmolysed against Ψ_e. Ψ_s, of the cell population at incipient plasmolysis is taken as the point of 50 per cent plasmolysis. Strictly this value for Ψ_s must be corrected for any turgor pressure by measuring the volume increase (Eq. [3.2]).

A direct measure of Ψ_s of vacuolar sap can be obtained from the extracted sap. If the cells are frozen, then thawed rapidly, ice crystals disrupt the membranes and the sap can be pressed or centrifuged from the debris. Ψ_s of the extracted sap can be measured conveniently with a freezing point osmometer. Another alternative procedure is to measure the water potential of air surrounding the tissue when it is sealed in a small chamber using a psychrometer; this measures Ψ of the intact cells. Next the tissue is frozen and thawed, to disrupt the protoplasts and liberate the vacuolar sap, and the procedure repeated. The psychrometer now measures Ψ_s of the sap. Unfortunately in all direct measurements of Ψ_s intracellular sap, which is mainly the vacuolar contents, is contaminated more or less by extracellular sap of different composition.

Turgor pressure component Ψ_p

We have seen in preceding sections that Ψ_p can be deduced by subtraction of Ψ from Ψ_s of cells or tissues. This method is the most commonly used in experimental work. (Sometimes Fig. 3.3 is transformed into a 'Hofler diagram' by rotating the lower part of the graph around the horizontal axis for $\Psi_{cell}=0$. This stresses the fact that at full turgor $\Psi_s=\Psi_p$ exactly, but it is confusing in making negative water potentials seem positive. Cell volume changes may also be plotted on the horizontal axis; this graph then confuses volume changes with Ψ_e quite unnecessarily.)

Turgor pressure of cells can be measured, in theory, by connecting a suitable pressure gauge to a cell vacuole by means of a suitable connecting probe. In practice it is too difficult to measure for routine purposes. The volume changes are normally very small and micromanipulation is required to insert the probes. Nevertheless Ψ_p has been measured in large algal cells (Green, 1968), sieve tubes (Hammel, 1968), guard cells (Edwards and Meidner, 1975) and epidermal cells (Steudle *et al.*, 1975). Turgor pressures are commonly surprisingly great – around 5 bar in parenchyma cells, 12 ± 5 bar in mesophyll cells and around 15 ± 5 bar in sieve tubes of well-watered plants. For comparative pressures see Appendix 4.

Negative wall pressures

In some cells there is evidence of negative wall pressures. For negative wall pressures to develop the cell lumen must be relatively narrow

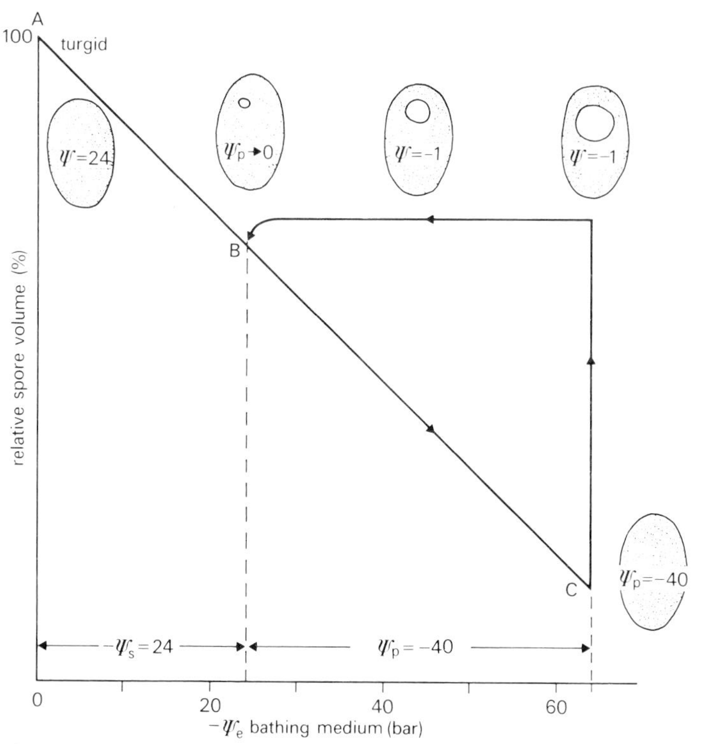

Fig. 3.4 The water relations of a *Sordaria* ascospore illustrating the influence of negative wall pressure. The water relations of the spore are regulated by the environmental water potential Ψ_e conveniently regulated osmotically. As water is extracted the cytoplasmic spore contents cavitate and an internal bubble (b) is produced in the ovoid spore.

and bounded by a strong or thick wall, so it seems most likely to occur in heavily lignified cells such as dead xylem conduits or woody fibres which may be living. Little is known about the development pressures in higher plants or the significance of negative wall pressure. It seems possible that if turgor pressures are zero or negative, cells may be more protected from sap sucking insects.

An interesting example of a cell which exhibits very large negative wall pressures is the ascospore of *Sordaria*. In these cells the volume enclosed by the wall changes very little and the protoplast and wall remain in contact at all times, even in an osmotic solution. In dry air the cell develops an internal bubble caused by cavitation of the cell contents. It is possible to measure the water potential necessary to induce cavitation using an external bathing medium, i.e. a salt solution; Ψ_e may be lower than -60 bar. If the water potential of Ψ_e is now raised progressively the internal bubble shrinks in size, eventually disappearing quite rapidly under the influence of surface tension. This stage corresponds to incipient plasmolysis because $\Psi_p = 0$. By inference the same spore in water must generate a turgor pressure Ψ_p which equals $+24$ bar by calculation (see Fig. 3.4), showing that Ψ_s is -24 bar. Since a negative water potential of -64 bar was required to initiate cavitation, the negative wall pressure, Ψ_p, generated in this cell was an amazing -40 bar!

In *Sordaria* ascospores, since there is no detectable change in the overall cell volume, the calculation of turgor pressures is greatly simplified in comparison with Fig. 3.3. Surprisingly *Sordaria* ascospores seem to survive and germinate after intracellular cavitation (see Milburn, 1970).

Measurement of other parameters

Turgor changes in cells are normally accompanied by elastic expansion or contraction of the cell wall. Such changes in cell volume reflect a storage capacity for water of both cells and tissues. The extent to which water is stored depends upon the elasticity of the cell wall. Stored water is an important water reserve during drought in some plants and in others it may act as a shock absorber, protecting the vascular system from excessive pressure changes which might cause cavitation.

Hydraulic capacitance

Most tissues live at water potentials less than zero and are composed of cells with elastic walls. Consequently when pieces of tissue are bathed in water, water uptake occurs and they increase in weight. This gives a useful measure of the extent to which a tissue is deficient of water, called the relative water content RWC, by Weatherley and others who developed it.

$$\text{Per cent RWC} = \frac{\text{Wt of tissue samples} - \text{dry wt}}{\text{Wt of tissue when fully turgid} - \text{dry wt}} \times \frac{100}{1} \qquad [3.3]$$

It does, in this form, equate with electronic capacitance, i.e. the capacity to store a quantity of electrons (amperes) under a given bombardment pressure (volts) in units called farads. To derive similar units for hydraulic capacitance we must relate per cent RWC to the water potential. Accordingly the hydraulic capacitance of a cell or tissue can be defined as the maximum mass of water it can absorb or release to become turgid expressed as a fraction of its water content when fully turgid over unit change in water potential. It can be measured in several different ways. One technique is to seal weighed pieces of turgid plant tissue in a series of chambers the water vapour in which is controlled by a range of osmotic solutions at known water potentials and at constant temperature. Eventually the pieces of tissue reach equilibrium with the chamber environment by gaining or losing

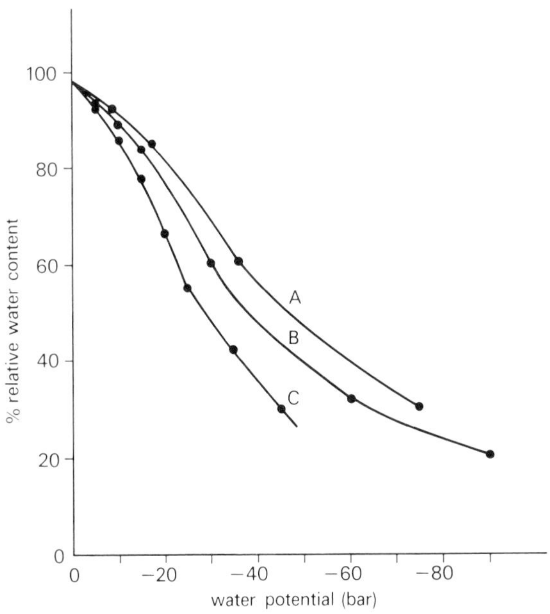

Fig. 3.5 The dependence of hydraulic capacity of leaf tissues on water potential. B Privet, C Tomato, (from Weatherley and Slatyer, 1957). A Brussels sprouts (from Milburn, unpublished).

water vapour. The tissues are then reweighed and the results plotted as per cent RWC against Ψ_e (see Fig. 3.5). The relationship is normally a sigmoid curve which corresponds with the lower curve in Fig. 3.3. Hydraulic capacitance data or curves such as Fig. 3.5 enable one to predict the amount of water which can be lost or absorbed by tissues during the onset or reversal of drought conditions. Alternatively hydraulic capacitance data can be used to compute the extent to which tissues or cells can behave as internal water reserves within plants.

The measurement of cell wall elasticity

If cell walls were inelastic the slightest reduction in water content would reduce the turgor pressure to zero or below zero. Since plants rely on the maintenance of turgor pressure for support, they would collapse by wilting. Most cell walls are in fact elastic which increases their capacity to store water and hence their resilience to wilting and also accounts for the 'sigmoid' curves shown in Figs. 3.3 and 3.5. Such behaviour can be detected if osmotica are applied to cells or a pressure bomb is used to express water from leaves or other plant organs.

Recently a successful technique has been described by Steudle, Zimmermann and Luttge (1977) to measure cell wall elasticity called, more strictly, the volumetric elastic modules ε which is a measure of the ability of a cell wall to resist pressure. To measure ε it is necessary to measure changes in external water potential Ψ_e accompanying small changes in volume which themselves cause small changes (0.1 bar) in pressure.

$$\varepsilon=\frac{V}{\Delta V}(\Psi_p-\Psi_s-\Psi_e) \qquad [3.4]$$

Volumetric elastic modulus	ε	MLT^{-1}	bar $\times 10^{-5}$ (=Pa)
Cell volume at full turgor	V	L^3	m^3
Hydrostatic pressure at full turgor	Ψ_p	MLT^{-1}	bar $\times 10^{-5}$ (=Pa)
Osmotic potential at full turgor	Ψ_s	MLT^{-1}	bar $\times 10^{-5}$ (=Pa)
Changes in above in response to Ψ_p changes	Δ	—	—

Zimmerman and others, having pioneered a new pressure transducer probe (see Fig. 8.4), found, in algal cells such as *Valonia*, maximum values of ε of 730 bar (Zimmermann and Steudle, 1975). They then successfully applied the same methods to bladder cells of the epidermis of *Mesembryanthemum* (a flowering plant). Assuming that the cells are spherical, as required by the formula above, ε was found to be 5 bar when the cells were totally relaxed and the turgor pressure Ψ_p was zero. As the cells approached full turgor, where Ψ_p was 3 to 4 bar, ε was found to increase to 100 bar. ε thus increased with increasing cell volume. Algal cells and *Mesembryanthemum* bladder cells seem to be typical of the plant world and for most plant cells ε is 0 to 100 bar (see Dainty, 1976).

Water transport through cells

During the ebb and flow of water through plants at some stages water must pass through cells. But what path does it take, via cell walls, or cell protoplasts, or both? How do organelles respond to changes in water availability and flow? To what extent are ions involved in cellular transport of water? These and similar problems have fascinated researchers of cellular conductivity. To answer the problems we must seek quantitative answers and the components must be studied and measured separately.

Cellular flow

Water flow through cells is inefficient when compared to vascular transport in requiring much greater pressure gradients to drive a comparable flow. This is reflected in plant anatomy. Cellular pathways which transport water in quantity (e.g. leaves) are seldom longer than five cells. How is this flow driven? If Ψ_s of the pathway cells is measured, by methods previously described, the results usually indicate that flow is osmotic, from higher to lower Ψ_s values. This may be a false indication because Ψ_p of the cells also plays a role. The overall driven gradient depends on the $\Psi_s + \Psi_p$ interaction which is the water potential gradient in the chain (catena) of cells. This is shown in Fig. 3.6 as a smooth gradient of Ψ, whereas Ψ_s data alone might seem to reverse part of the actual flow.

In this section we will concentrate on protoplast, membrane and organelle conduction. Cell wall conduction is discussed in Chapters 4–6. Measurements of water flow through cells have concentrated on protoplast conduction and two different measurements have been made – pressure flow and diffusional flow. The former relates to water

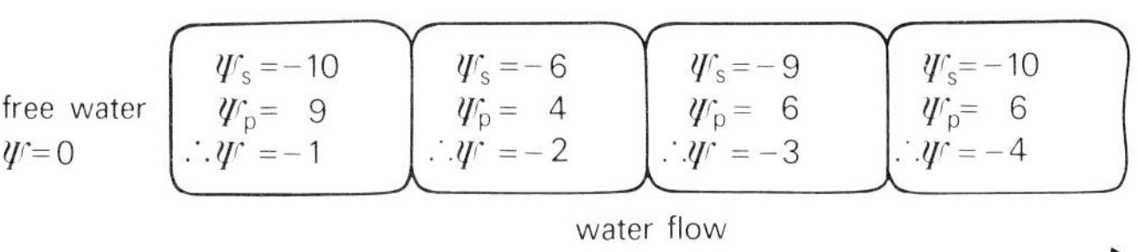

Fig. 3.6 Schematic diagram to show how water could flow through a catena of living cells along a water potential difference $\Delta\Psi$, of 4 bar which does not necessarily coincide with Ψ_s or Ψ_p gradients.

flow through cells along a water potential gradient, the latter to the exchange of water between a cell and the flow past it.

Pressure flow It is usually impractical to induce pressure flow through cells hydrostatically, so in practice osmotic solutions are used to drive water through cells at equal hydrostatic pressure relying on the difference in internal pressures of the osmotic solutions to drive the flow. The simplest method is to measure the volume of the protoplast within a plasmolysed cell microscopically. The water potential of the bathing solutions is then altered to generate a water potential difference across the protoplast membranes. The volume change per unit time is measured across the mean surface of the protoplast which represents the changing cell membrane area. Using the following equation the hydraulic conductance of the plasmalemma can be found on the assumption, which is probably correct, that the plasmalemma, and neither the tonoplast nor the cytoplasm, is the most significant resistance to flow.

$$J_v = \frac{\Delta V}{At} = L_p(\Delta\Psi) \qquad [3.5]$$

Specific volume flux	J_v	$L^3L^{-2}T^{-1}$	$m^3\,m^{-2}\,s^{-1}$ $(=m\,s^{-1})$
Volume change of protoplast	ΔV	L^3	m^3
Area membrane (mean)	$A=\frac{1}{2}(A_1+A_2)$	L^2	m^2
Time	t	T	s
Hydraulic conductance	L_p	$LT^{-1}(ML^{-1}T^{-2})^{-1}$	$m(Pa)^{-1}\,s^{-1}$ $(=m\,bar\,s^{-1}) \times 10^{-5}$
Water potential difference	$\Delta\Psi$	$ML^{-1}T^{-2}$	Pa$(=bar \times 10^{-5})$

This method suffers from a number of defects. To begin with the plasmalemma has been changed because plasmodesmata have been pulled away from cell walls, broken and then repaired, over a good deal of the plasmalemma surface. Another serious objection is that osmotic solutions both within and outside the cell fail to act at the initial concentration at the membrane surfaces on account of the development of local diffusion (or unstirred) layers of solution adjacent to the membranes. This latter problem can be partially overcome by using protoplasts from cell walls. The membrane area and thickness also change during the measurement. A technique was introduced designed to overcome some of these problems called *transcellular osmosis*. In this case an osmotic gradient causes the influx and efflux of water across a large whole cell, e.g. a giant algal cell which is sealed between two chambers containing different solution. The flow of water through the cell is measured on capillary potometers (Fig. 3.8). Numerous complications arise in interpreting these results. By this method, cellular permeability, hence J_v, seems

Fig. 3.7 One representation of membrane structure showing hydrophobic lipids with polar groups outermost forming a matrix containing proteinaceous materials permeable to water and supporting ionic pumps, enzymes, sensory structures, etc. Any *static* representation of a membrane is liable to be illusory (from Sjostrand, 1968).

much higher than is indicated from plasmolytic studies, possibly because the hydraulic conductivity of the cell wall under the 'seal' is included in the measurement. Equation [3.5] is used except that ΔV is measured directly using the capillary potometers (see Fig. 3.8), A and $\Delta\Psi$ remain essential constant. Flow is across two sets of membranes. The problem of diffusional layers remains and may become very serious if the flow rate is high (see Fig. 3.9, for example). Predictably the cell exhibits asymmetric flow properties.

Diffusional flow Another technique used widely to measure permeability in recent years utilises the isotopic exchange of water (e.g. tritiated water) or solute molecules across cell membranes. This

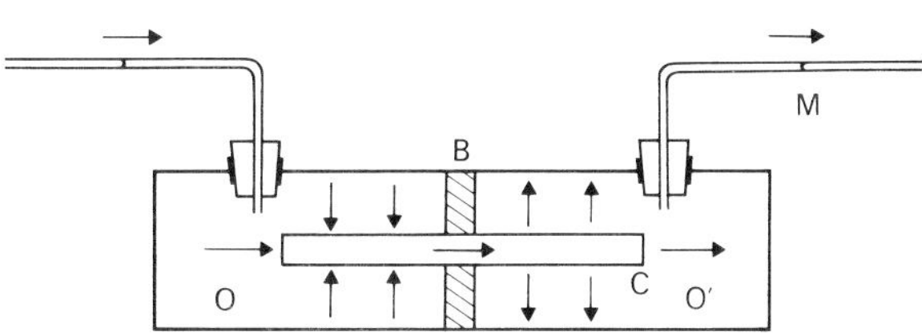

Fig. 3.8 Schematic representation of a potometer to measure transcellular osmosis in a cell C sealed in a barrier B. Flow generated by the osmotic gradient $\Delta\Psi_s/x$ O to O′ is measured by movement of the potometer meniscus M.

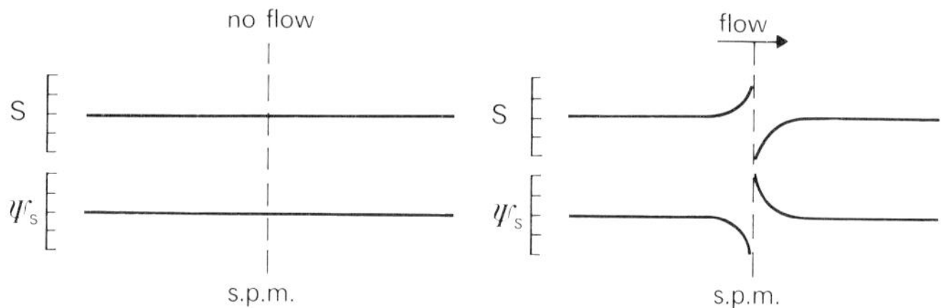

Fig. 3.9 Pressure flow can induce an apparent decrease in conductivity as the flow rate increases because solutes S tend to accumulate on one side of a membrane but are depleted from the other. The effect establishes an osmotic gradient $\Delta\Psi_s/x$ opposing flow.

method has the advantage that there is little interference with cell turgor. In practice rather than measure cell loading, it is simpler to load a cell or tissue with an isotope and then study the rate of efflux (unloading) by transferring it at regular intervals to an elution series which is then assayed for radioactivity. The method is also attractive in that the cell is subject to very little manipulation. It is important to realise that the diffusional coefficient determined by this method does *not* equate with the hydraulic conductance determined by pressure flow methods. The reason for this is that a term must be included in

Table 3.1 Examples of L_p, the hydraulic conductance, determined for membranes of higher plant cells and giant algal cells representing the range of values determined using plasmolysis P and transcellular osmosis T techniques. It has been assumed that within experimental error 1 atm = 1 bar. For comparative purposes L, the hydraulic conductivity, has been calculated assuming a membrane thickness (path length x) of 7 nm ($L = L_p \times x$).

Method	Plant	$L_p \times 10^{-9}$ (m bar^{-1} s^{-1})	Worker	$L \times 10^{-18}$ (m^2 bar^{-1} s^{-1})
P	*Beta vulgaris* (not plasmolysed)	0.2	Myers (1951)	1.4
P	*Beta vulgaris* (when plasmolysed)	3.6	Myers (1951)	25.2
T	*Valonia ventricosa*	1.8	Gutknecht (1968)	12.6
T	*Chara corallina*	101	Dainty and Ginzburg (1964)	707
T	*Nitella transluscens*	107	Dainty and Ginzburg (1964)	749
T	*Nitella flexilis*	120	Kamiya and Tazawa (1956)	840
T	*Nitella flexilis*	300	Kamiya and Tazawa (1956)	2,100

the equation to deal with frictional interactions between the marker isotope and water which increases the 'unstirred layer' effect. Conventionally the diffusional coefficient has been derived from Fick's law omitting this term, which probably explains why most diffusional coefficients are lower than those derived from pressure flow driven by osmosis; see examples in Table 3.1.

The measurement of membrane permeability

There is controversy over the correct concept of membrane structure between those who would consider it as a barrier to diffusional flow and others who would regard it as a barrier traversed by 'pores' of various dimensions. Many thermodynamic treatments are based on the pore model, which is the more easy to understand conceptually.

Functionally we must explain how membranes can transport lipophilic substances, water hydrophilic solutes (e.g. sucrose), and ions. Some ions are undoubtedly pumped against their electrochemical gradient by electrogenic pumps which must be located in the membrane; others merely diffuse down an electrochemical gradient. Other pumps, specific for certain solutes, such as sucrose, must also be located within the membrane. It is also reasonably clear that receptors for gravitational light and pressure stimuli are also located within plasmalemma membranes because they are the only part of the cytoplasmic apparatus sufficiently stable to support them (Fig. 3.7). Membranes alter their properties in response to changes in their environment ranging from temperature changes to hormonal actions and any comprehensive model must enable us to explain these changes.

There seems little doubt from studies on the permeation of different molecules and ions that membranes behave as if they contained long (above 7 nm) water-filled pores, sufficiently narrow (say, 0.4–1.5 nm) to have a selective effect on non-electrolytes and hydrated ions (see Table 3.2). A problem arises if the walls of the pores are hydrophilic because we would expect an ice-like water to be held by hydrogen bonds as a lining up to 1 nm thick. On this basis, since the penetration of water molecules must be allowed for in our model and a water molecule is about 0.3 nm diameter, then structurally the pore

Table 3.2 Reflection coefficients for some solutes with reference to the giant algal cell *Nitella flexilis* (Steudle and Zimmermann, 1974). For other plant cells the coefficients form a similar sequence but have different magnitudes.

Solute	Sucrose	Glucose	Glycerol	Isopropanol	Ethanol	Methanol
Reflection coefficient σ	0.97	0.96	0.80	0.35	0.34	0.31

must form a tube with an outermost diameter of at least 2.3 nm. If, on the other hand, we suppose the pore to have a more hydrophobic lining, i.e. between protein–lipid or lipid–lipid molecules it need be a mere 0.3 nm or so diameter, but should repel water molecules! One possibility is that the above model is unrealistically simple. It is likely that biological membranes have ephemeral 'pores', which are constantly being formed and blocked, even along their length. Hopefully new techniques of X-ray interferometry now capable of high resolution to 0.5 nm may solve the problem soon. Interest has focused on membrane structure and function for a better understanding of ion and non-electrolyte transport in addition to water. In this context, it has proved useful to isolate membranes as far as possible. The hydraulic conductance L_p of a membrane is generally a more useful measure of conduction (because it is not specific as to the dimensions of the pathway) than L the hydraulic conductivity. To calculate the latter for a membrane we need to know its thickness x (usually about 7.5 nm) when $L_p \times x = L$.

Water transport in ultrastructural organelles

Optical microscopy of fixed (dead) plant material gives a false impression of simplicity in membrane systems and cellular organelles. Observations of cytoplasmic activity in living cells and electron microscopy have combined in recent years to compel recognition of the dynamic turnover of materials and membrane systems and the diversity of organelles in the cytoplasm.

The study of the physiology of these organelles is still at an early stage but two basic techniques are well established. Plastids or mitochondria are removed from cells by mechanical disintegration. The chosen organelles are then separated from the debris by high speed centrifugation in media of appropriate density. Volume changes in the suspension of organelles can be monitored conveniently by measuring changes in optical density, a technique developed from studies of blood erythrocytes. In this way Nobel (1974) was able to show that disc-like pea chloroplasts decreased in volume in response to a decreasing water potential of the bathing medium.

Slowly the electron microscope studies of many static ultra-thin sections has allowed the building of a complex three-dimensional, dynamic picture. Most membranes of a cell, excluding those surrounding such structures as chloroplasts, can be regarded as part of an endomembrane which includes tonoplast and plasmalemma, but also rough and smooth endoplasmic reticulum, outer nuclear membrane, secretory vesicles and dictyosomes (Golgi apparatus). If a short pulse-label of a radioactive amino acid is applied to tissues, its incorporation and transfer stages can be followed autoradiographically in whole cells or in organelles after fractionation, as described above.

These studies have shown that membranes of both plant and animal cells are continually being turned over and that materials, including water, are transported in the form of vesicles, for example during the growth of cell walls. The frequency and speed of vesicular transport is staggering. These wall materials are transported in vesicles for tip growth of pollen tubes. Vesicles were produced by the dictyosomes at a rate of 16 vesicles s^{-1} to maintain the steady growth in tube length observed (10 nm s^{-1}) (see Morre and Mollenhauer, 1974). Such vesicle transport may represent a rather weak form of active water transport, since water is transported by metabolic activity without obvious recourse to ion movement and diffusional processes, such as osmosis. How plants may drive such a mechanism remains to be seen.

Transfer cells

It is certain that what have been named 'transfer' cells by Gunning and Pate (1969) have been observed by botanists in a number of different plants for many years (e.g. Haberlandt, 1914). The fine tuft-like processes which are extensions of the primary cell wall are found deposited on specific areas of cell wall. They can easily be seen in thin handcut sections of, for example, *Tradescantia* nodes stained in toluidene blue. They appear as a dense fibrous layer orientated radially, apparently within the plasmalemma. Optical investigations have been greatly complemented by electron microscopy (EM). Attention has been focused on the fact that transfer cells are not 'new' cells but constitute a modification of standard cell types (epidermal xylem parenchyma, companion cells, etc.) to fulfil a physiological function.

Transfer cells seem to be directly involved in cellular transport of solutes between cells over short distances. This was inferred because transfer cells are located in many strategic situations such as in the cells surrounding minor veins in leaves, in placental tissues and especially in nodal tissues between xylem and phloem conducting elements. The wall area is increased locally up to twenty times by the finger-like ingrowths of primary wall forming a wall-membrane apparatus. The ingrowths are more permeable than normal secondary walls, as shown by the penetration of lanthanum hydroxide (an EM stain). Their role in transport is also inferred from the telling observation that their morphogenesis is reduced when solute fluxes are limited through prevention of assimilation by reduced light or carbon dioxide starvation.

Transfer cells are probably analogous with microvilli in animal cells in gall bladder, intestine and kidney, which are known to implement transport of solutes. This transport function can occur in both organs of secretion and absorption and may involve a mass flow generated by an osmotic gradient as in the *perpetuum mobile* machine (Ch. 2).

Whether the driving mechanisms are electro-osmosis or vesicle transport remains to be established (see Gunning and Pate, 1974). Since transfer cells probably indicate 'bottlenecks' in long distance sap transport systems, fascinating questions are raised both by their presence, and even more by their absence in comparable plants.

Effects of electrical potentials

It is important to consider the effects of electrical potentials when studying water flow. Electrical potentials can profoundly affect ion transport and even drive ions against a concentration gradient. Ion accumulation has osmotic effects which can modify cell turgor (Ψ_p). From the extensive literature on ion transport we will consider a useful test for passive ion movement, Mitchell's chemiosmotic hypothesis which now plays an important part in relating electron transfer, ATP metabolism with osmotic effects and also electro-osmosis which may influence water flow directly at a cellular level.

The Nernst equation

Most plant cell membranes are found to be electrically charged when studied with microelectrodes and a high impedance voltmeter (a high resistance is necessary to stop the system discharging). Electrical gradients are involved in ionic transport phenomena through the membrane which in turn controls water flow through osmosis. Ion flows (fluxes) may be passive (when they tend to obey Fick's diffusion laws), or actively pumped. It is important to recognise that they only tend to follow concentration gradients passively, because the really important factor is the electrochemical gradient, which is the resultant of concentration and electron density gradients. The important consequence is that an electrical gradient can cause *passive* accumulation of an ion *against its concentration gradient.*

For any given ion the electrical potential for a passive flux giving rise to ion accumulation can be calculated by applying the Nernst equation (but see App. 14*c*):

Nernst equation $$E_N = \frac{-RT}{zF}\ln\frac{c_o}{c_i} \qquad [3.6]$$

Term	Symbol	Dimensions	Units
Electrical potential gradient	E_N	–	mV
Valency	z	–	–
Universal gas content	R	8.3143	$J\,mol^{-1}\,K$
Absolute temperature	T	–	K
Faraday	F	96,487	$C\,mol^{-1}$ (or $J\,mol^{-1}V^{-1}$)
Natural log	ln	–	–
Concentration inside cell	c_i	ML^{-3}	$mol\,m^{-3}$
Concentration outside cell	c_o	ML^{-3}	$mol\,m^{-3}$

If the electrical potential E_N predicted for a given ion from the Nernst equation differs significantly from that observed (E_{obs}) there is good reason to suppose that the ion is being actively transported against its electrochemical gradient (see Table 3.3). This can in principle lead to ionic concentration gradients being set up which can

Table 3.3 Ions in extracted sap (mainly vacuolar in origin) which, with organic anions, account for the osmotic potentials of vacuolar sap in leaves and shoots of higher plants (from Eaton, 1942). Under different conditions the extracted sap composition would also be different, even from the same species, and might include significant quantities of organic acids and sugars.

Genus and species	Ionic composition (mM)						Osmolarity
	K^+	Na^+	Ca^{2+}	Mg^{2+}	Cl^-	SO_4^{2-}	Osmol m^{-3}
Lycopersicum esc. (Tomato)	161	6	58	42	25	75	425
Gossypium hirs. (Cotton)	228	18	126	54	18	92	620
Beta vulg. (Beet)	167	229	1	51	44	20	610

produce osmotic transport of water despite the absence of a Ψ gradient overall. Metabolic energy (ATP) is required to maintain such transport.

Mitchell's chemiosmotic hypothesis

In 1966 Mitchell introduced the idea that the link between metabolically driven electron transport and phosphorylation was a proton gradient. His modified chemiosmotic theory is widely accepted and assumed to operate in many systems in plants and animals. It links the capacity of membranes to develop electrical potentials with their capacity to regulate ATP (adenosine triphosphate) synthesis and hydrolysis. Proton pumps must be located in membranes to be effective. In chloroplasts protons are driven by electron flow under the influence of light (photosynthesis). This establishes both a pH gradient and an electrical potential across the thylakoid membrane. The theory states that if ATP is hydrolysed also protons (H^+) are pumped in the same direction as electron transport. Conversely if H^+ flow is reversed ATP is synthesised. This chemiosmotic mechanism is believed to be operative in mitochondria and chloroplasts (see Fig. 3.10). Strong evidence for the chemiosmotic hypothesis has been obtained, for example by Jagendorff who created a proton gradient across chloroplast membranes and this in turn caused ATP synthesis even in darkness.

In light chloroplasts pump protons which usually makes them swell osmotically. Osmotic potentials depend on the relative numbers of small particles on different sides of a membrane. The proton is itself osmotically neutral, producing equal numbers of protons and hydroxyl ions on each side of the membrane. ATP hydrolysis does produce an osmotic change but is too dilute to be significant. So what is the link between the swelling of chloroplasts (or guard cells) and proton pumping driven by light? Apparently this depends on the movement of other solutes (cations, anions or neutral solutes such as sugars). Coupled active flow of these solutes occurs down the

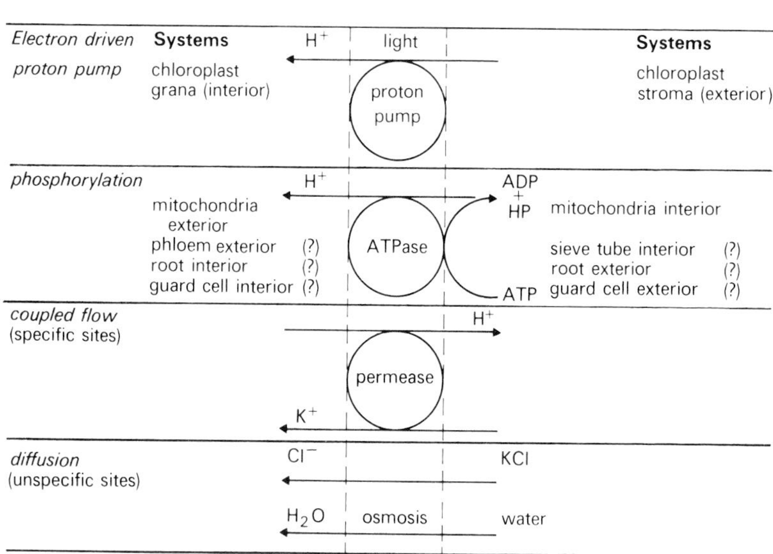

Fig. 3.10 Simplified diagram showing how light or ATP hydrolysis could drive proton fluxes across a membrane in accordance with the chain-osmotic hypothesis. The proton gradient drives KCl to the left leading to an osmotic flow of water. Under different conditions, for example if there was an anion-hydroxyl coupled permease in the membrane Cl^- would be moved right and if K^+ could flow passively, KCl would move to the right. Net osmotic potential changes depend on permeability characteristics of the membrane system.

electrochemical gradient via specific systems such as permeases which are powered by ATP or proton flow. Diffusion of solutes and water also takes place as an unspecific 'leakage' across membranes depending on their permeability characteristics. In this way proton fluxes can cause accumulation of cations and anions, as shown in Fig. 3.10. The net effect is to decrease intracellular Ψ_s causing water also to diffuse into an organelle system by osmosis making it swell or to become more turgid.

Electro-osmosis

Water flow can be driven directly under the influence of an electrical gradient acting across a charged porous system, e.g. cell walls, membranes, etc., by the process of *electro-osmosis.* A pore lined with fixed negatively charged surfaces repels, for example, chloride anions from potassium chloride solution but allows the potassium cations to migrate through the 'centre' of the pore. Cations, such as potassium, when attracted along an electrical gradient towards the negatively charged end of the pore, draw water molecules with them also by frictional drag producing a flow of water. This separation of charged ions modifies the electrical gradient eventually reaching equilibrium in accordance with the Nernst equation. The acidity–alkalinity (pH) of the medium is modified also because weakly charged H^+ ions remain with the Cl^- anions, while weakly charged OH^+ ions are transferred across the membrane with potassium cations. For flow to continue the ions must be kept in motion.

For many years it has been proposed that electro-osmosis plays an important role in long distance transport systems of plants (such as phloem) but as yet evidence is not convincing. On a cellular and subcellular scale the role of electro-osmosis may be more important but again firm proof is lacking. The presence of electrical gradients in producing pressure differences may be calculated to be remarkably large (see Nobel, 1974):

$$E = 1\,\mathrm{mV} \equiv 32\ \mathrm{bar} = \Delta\Psi_p \qquad [3.7]$$

So perhaps, since E may often reach $-150\,\mathrm{mV}$, we should keep an open mind on the topic. Electro-osmosis may turn out to be more important than we now suppose!

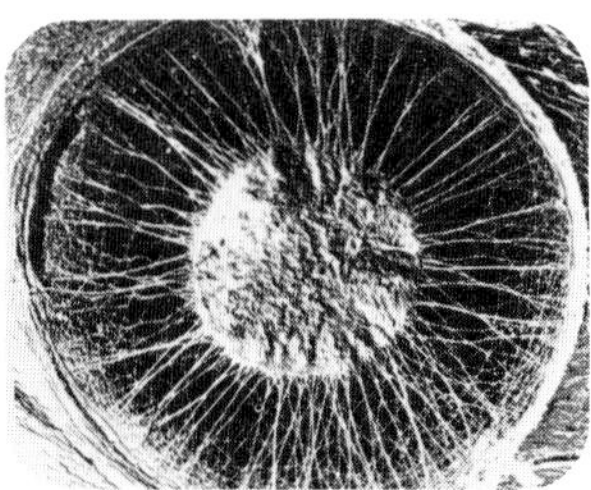

Chapter 4

Roots and uptake from soil

When plants first began to invade land, soils must have been hostile substrates indeed. Gradually the cyclic decomposition of organic materials substantially modified this primitive structure making it a less hostile medium from which the plant derives support, a supply of water and a source of inorganic nutrients.

Properties of soil

Physically most soils can be separated, somewhat arbitrarily, into four main materials by shaking a sample in water to form a soil suspension.

1. *Sands* and other coarse particles sediment rapidly and have large spaces between compacted particles.
2. *Silts* are composed of finer particles which settle from suspension more slowly.
3. *Clays* are even finer particles which settle from suspension very slowly and have only minute spaces between compacted particles.
4. *Humus* consists of dark coloured organic debris in the process of decomposition making up 2 to 100 per cent (in peats) of the soil fraction. Unlike the three previous grades which are considerably denser than water (about three times), humus has an apparent density near unity and often floats, on account of entrapped air, in a soil suspension.

The state of division of the particles in soil strongly affects the soil's capacity to retain water. Generally the particles attract water molecules which are then retained through surface tension, as if the spaces between particles were small irregular capillary tubes. The smaller the particles are, the greater the effect of surface attraction, and the interparticulate pores become narrower. Pores in sands are predominantly 30 μm or more, in silts 0.2–30 μm and in clays less than 0.2 μm. Conventionally the physical attraction of the soil matrix by surface tension and imbibitional or adhesional forces are grouped together to give a *matric potential* Ψ_m. An additional factor is sometimes added to incorporate the effect of gravity, the gravitational potential, Ψ_g, but this is often insignificant and can be ignored. In fact there is little justification to distinguish these parameters from our familiar Ψ_p, the pressure potential.

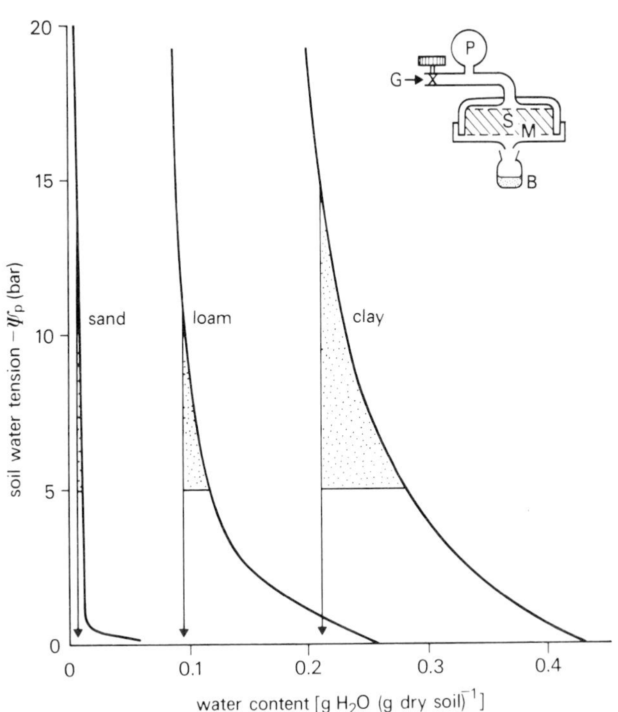

Fig. 4.1 Soil characteristic curves for sand loam and clay soils. For a soil water table at 3 m depth $\Psi_p = 0.25$ bar. Between this tension and, say, −5 bar, to the limits of water extraction by roots (the permanent wilting point), around −15 bar, lies the useful hydraulic capacitance or water reservoir which increases sand < loam < clay. *Inset:* Diagram of soil pressure-plate apparatus containing soil S on membrane M at pressure *P* generated by gas supply G which drives water into the bottle B.

The particulate yet porous nature of soil affects its pressure potential Ψ_p. Coarse particles in sand retain water only weakly, but finer particles in clays retain water strongly. (For example montmorillonite clay particles adsorb up to ten water shells each 0.4 nm thick during saturation.) These soil characteristics are expressed in curves (Fig. 4.1) where as the proportion of water is reduced, the pressure (or matric) potential decreases rapidly and exponentially. A reference point used in work on soils is the *field capacity*, defined as the quantity of water retained in field experiments 24 hours after a thorough soaking. In laboratory experiments we may define it as the capacity of fully saturated soil when gravity has removed excess water. Of course, if a soil is disturbed, capillary spaces are formed or reduced and its field capacity is altered. So although field capacity corresponds to the fully saturated condition as a reference standard (cf. relative water content for plant tissues), it is not wholly satisfactory. To plants the really important criterion is the *hydraulic capacity* of soils between field capacity and about −15 bar water potential the '*permanent wilting point*' beyond which water ceases to be available to them (see Fig. 4.1).

A further important consequence of the porous nature of soil is that of *hysteresis*. This means that the relationship between the matric potential and the water content is different if a soil is being dried or moistened. The reason for this is that the small irregular pores in soil attract and hold water during drying until finally they are emptied as the soil shrinks and dries (Fig. 4.2). Having dried, water has difficulty

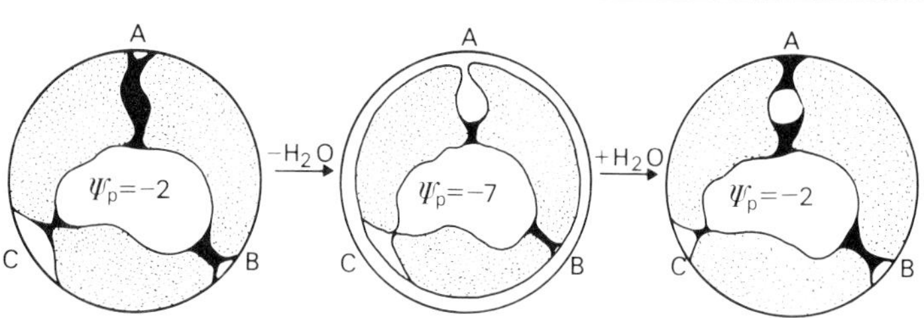

Fig. 4.2 Diagrammatic representation of hysteresis in soils. Water held at $\Psi_p = -2$ bar is removed as the soil dries and shrinks to $\Psi_p = -7$ bar. The transport channels narrow reducing hydraulic conductivity. On rewatering to $\Psi_p = -2$ bar air trapped in the system reduces the hydraulic capacity below the initial value.

refilling the tubes on account of entrapped gas bubbles in the shrunken capillaries so reducing the relative water content of the soil in comparison with the same soil at the same water potential during drying.

So far we have considered the relation between the capacity to retain water and the pressure potential, but a factor vital to plants is the ability of the soil to transport water, its *hydraulic conductivity*. The hydraulic conductivity is high when the soil approaches saturation with water, but as the effective capillary pores are progressively eliminated by emboli of air the hydraulic conductivity is drastically reduced in dry soil, and this effect persists on account of *hysteresis*, even when the soils are rewetted. The important consequence for a plant, liable to suffer from water shortage in drying soil, is that surface contact with soil must be maximised so that the conducting pathway in soil is of minimum length over a maximum area.

Chemically the commonest component of soils is silica (quartz), often in the form of silicates, but soils are extremely heterogeneous with many feldspars, granites and aluminates present. Due to the action of mechanical disturbance, such as frost, expansion effects and erosion by chemicals, soils are continually under degradation so that new ions are liberated, though they may be leached away by rainwater. Humus also liberates a further supply of inorganic ions depending on its rate of decomposition. The presence of carbonates and bicarbonates is in part caused by respiration of soil micro-organisms and plant roots. In recent years it has become clear that quite significant proportions of ions are leached directly from plant leaves by rainwater and that many other substances are released by micro-organisms, including antibiotics, organic acids and mucilages which may influence the ecology around roots in the *rhizosphere*.

In moist soils the water potential is reduced by the solutes, so that Ψ_s equals about -1 bar. However, in saline soils Ψ_s can decrease to about -22 bar, the value of seawater, or even lower if the soil dries. In general if the volume of soil water is halved Ψ_s is doubled, but this depends on the solubility of the components. Naturally if fertilisers are added the osmotic potential of soil water may decrease. To some extent roots are able to digest solid materials, as can be shown by growing roots over tiles of marble, etc. After a period the tiles are found to be etched where roots have made contact. The cause of this etching seems to be a release of organic and carbonic acids by roots, but the micro-organisms of the rhizosphere are influential also.

The measurement and control of water potentials of soils

The measurement of soil water potential, Ψ_{soil}, is rendered very difficult by the soils, low hydraulic conductivity when dry, which leads to considerable heterogeneity, and the fact that its water potential may be changed by disturbance (see above). In principle,

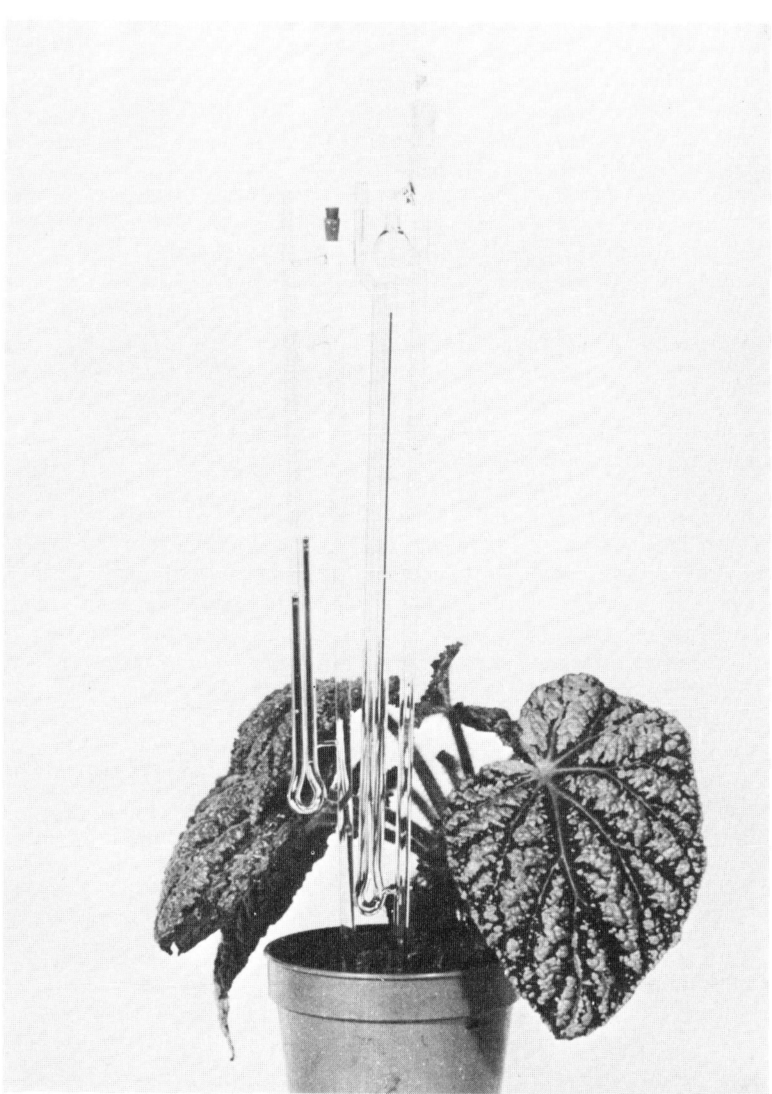

Fig. 4.3 The course of tensions registered on a widebore soil tensiometer (as used in Fig. 4.4) and a narrow bore tensiometer. Tensions differ because soil around the porous clay probes embedded in the same pot of soil is wetted as water is released in greater amounts from the widebore system than the narrowbore system.

therefore, embedded probes are to be preferred to samples removed from the soil mass, unless special precautions are taken to avoid compaction or loosening. Since the osmotic potentials are reasonably small and correlated with pressure potentials most techniques measure the latter.

Soil tensiometers

If a porous pot, filled with water and sealed to a pressure gauge, is embedded in moist soil, as the soil dries it extracts water from the apparatus so that a negative pressure is registered. As the soil continues to dry the pressure potential Ψ_p continues to fall (Fig. 4.3) until at around -0.8 bar the tensiometer invariably suffers an embolism as water sucks air through the pores in the pot (see Fig. 2.3). It would seem therefore that a soil tensiometer ought to measure pressure potentials in soils to -0.8 bar only, which would severely limit its usefulness. In fact, however, the gauge in Fig. 4.4 registered only a fraction (say -0.5 bar) of the negative pressure developed, on account of the fact that water was withdrawn from the pressure gauge as it registered, moistening the soil locally around the porous porcelain probe. This moist layer also protects the tensiometer from embolism, caused by the pores of the probe drying out even though the surrounding soils may be quite dry; the reduced hydraulic conductivity of soil contributes to this heterogeneity in the system. In practice this extremely simple and convenient technique is best calibrated against another technique based on sampling.

Neutron scattering

Fast neutrons are slowed considerably by H atoms and the main source of these is in soil water. A probe can be lowered down an access tube with a suitable neutron source positioned near it and the reflected neutrons reaching the detector probe are counted on the soil surface using a scaler. Though this technique might seem ideal for soil moisture measurement, it suffers from several sources of error including effects of hydrogen atoms in organic matter and by the scattering from other atoms, especially chlorine.

Electrical resistance blocks

In this technique, developed by Bouyoucos, stainless steel grid electrodes are embedded in plaster of paris (gypsum) blocks which, in turn, are embedded in soil. As the blocks release water to the soil their electrical resistance increases until the soils are soaked, when their water content and resistance fall again. This technique is very convenient, requiring only a portable resistance bridge carried from probe to probe, but it is difficult to calibrate in absolute units and suffers from *hysteresis* problems.

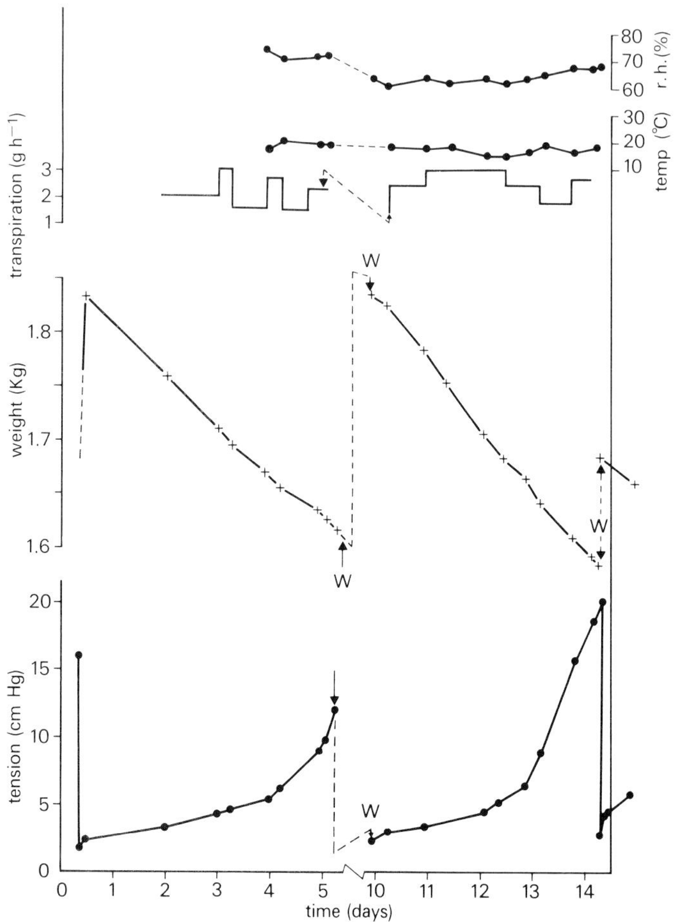

Fig. 4.4 The course of soil tensiometer readings from an experiment on a potted *Pelargonium* plant. Transpiration determined by weighing the whole system remained remarkably constant despite clear fluctuations in soil water tension which rises (i.e. Ψ_p falls) until the pot is watered W with varying amounts of water. The tensiometer only indicates relative Ψ_p of the soil as it dries, but probably gives an accurate measure when the soil is wet. It never becomes zero because of the effect of gravity on the soil water.

Pressure plate apparatus

In principle the pressure plate apparatus is similar to that of the Scholander pressure bomb. A soil sample, supported on a pressure plate is subjected to compressed air and the water expressed at different pressures is used to construct a soil moisture curve (see inset Fig. 4.1). The geometry of the soil chamber is important and errors can arise from abnormal gas flow and water flow through the specimen.

Other methods

Several other methods can be utilised to measure soil water potential. The number of methods indicates the lack of ideal techniques (rapid, inexpensive and accurate). In general, soils are calibrated by weight, the relative water content being related to dry weight (105°C) or, better, the moisture content at saturation. Cryoscopic and psychrometric methods have been used to determine pressure and solute potentials. An interesting technique for their independent determination is to use a centrifuge to expel soil water under a known centrifugal pressure and then measure its osmotic potential using an osmometer.

The extraction of water from soils

It should not be assumed that roots are solely concerned with water uptake; roots serve many functions including anchorage and support and are specialised for nutrient uptake, symbiosis (as in mycorrhizas), water uptake and even gas exchanges (as in mangroves). The role of roots in filtering soil water has, perhaps unwisely, been neglected. Filtration systems of comparable efficiency become choked rapidly in laboratory experiments.

Soil hydraulic conductivity, L_{soil}

It is possible to measure the hydraulic conductivity of a cylindrical plug of soil on the basis of Darcy's law as shown in Chapter 1, where:

$$J_v = L_{soil}\frac{\Psi_p}{x}. \qquad [4.1]$$

The hydraulic conductivity of soil varies by many orders of magnitude as soil dries. It is a complex process depending on both liquid conduction and distillation. In nature, soils fluctuate in temperature causing expansion and contraction of particles and gaseous contents. Typical hydraulic conductivity measurements are listed in Table 4.1.

Gross structure

Shallow rooted plants are poorly situated to survive drought but ideally placed to collect nutrients from the richer soil horizons near

Table 4.1 Hydraulic conductivity of different soil types at differing degrees of moisture saturation expressed in terms of approximate water potentials Ψ_p. The radius of curvature of stable water menisci between the soil particles is indicated also. Surface soil is often unrepresentative of deeper soil, being much drier and an effective insulation. Its hydraulic conductivity effectively equates with transport via the gas-phase when it is dry (see Tables 6.3–6.5).

Soil type	Water potential (pressure) Ψ_p bar	Meniscus radius of curvature (water films) $10^{-6} \times$ m (= 1 μm) (20°C)	Hydraulic conductivity L $10^{-4} \times m^2\,bar^{-1}\,s^{-1}$ (= 1 $cm^2\,bar^{-1}\,s^{-1}$)
Sand (wet)	−0.1	14.5	1–10
Sand (dry)	−15	9.5	10^{-8}
Clay (wet)	−0.1	14.5	10^{-4}
Clay (dry)	−15	9.5	10^{-8}
Surface (dry)	−15 to −1,000	–	10^{-7}–10^{-10}

the surface. Deep tap roots are better adapted for survival during drought being able to collect water from deeper soil horizons, which are protected from evaporation and can act as important reserves. The *water table*, the depth at which soil is saturated with water, normally 'dives' during drought. By examining the roots of a plant it is often possible to deduce its strategy for survival both for an individual plant or species, or genera type. In addition to gross morphological distribution, root hairs improve contact with soils by greatly increasing the total surface area of roots.

Root hairs

Not all roots possess root hairs and some produce root hairs in dry soil but not in wet soils (*Citrus* spp). This observation suggests that root hairs may be a special device to make more intimate contact with soil particles. Since the cell walls of root hairs provide an excellent path for water transport, in strong contrast with a dry soil, they greatly enhance the capacity to absorb water in times of drought. Root hairs are formed from epidermal cells of roots. In moist air they usually produce long regular sausage-like outgrowths (the standard textbook type!), but in soils they are quite different being irregular and closely attached to soil particles, partially on account of their turgor pressure and partially because of hydrogen bonding, which is why small seedlings pulled out of soil have considerable quantities of soil attached to their roots. The intimate contact with soils probably functions mainly for water uptake, but undoubtedly nutrient absorption would be improved, particularly where digestion of soil particles is possible. When plant roots suffer from a restricted water supply they tend to lose turgor and shrink (reaching in some cases 25 per cent of the former diameter) away from the soil mass leaving an air gap (perirhizal drying zone). Root hairs probably minimise this tendency by maintaining intimate contact.

Root structure and water uptake

Water moves freely through cell walls towards the stele. A certain fraction may be transmitted through protoplasts, but the hydraulic conductivity of protoplasts seems too high for this fraction to be more than about 10 per cent. There can be no doubt that water penetrates freely via cortical cell walls to the endodermis. A long time ago Rufz de Lavison (1910) showed that solutions such as ferrous sulphate, fixed *in situ* with potassium ferricyanide as prussian blue, reached the Casparian strips in the endodermis. Similar results have been obtained recently by Tanton and Crowdy (1972) using lead chelate fixed by hydrogen sulphide gas to give insoluble lead sulphide and from studies on the uptake of uranyl acetate (Robards and Robb, 1972) studied by electron microscopy. Other solutes, e.g. ammonium salts, are not halted by the endodermis, probably because they are transported in the symplast as described below. In older roots the endodermis develops an inner suberin layer and then a thick cellulose lining rendering the system almost impermeable.

The Casparian strips serve as a 'damp-course' (Fig. 4.5) preventing the *free* access or escape of water from the stelar tissues of roots or even subterranean stems. In transverse sections, by microscopy, the Casparian strip seems insignificant. By far the most satisfactory way to appreciate its extent is to digest a root in concentrated sulphuric acid; only the Casparian strip remains from the cylinder of

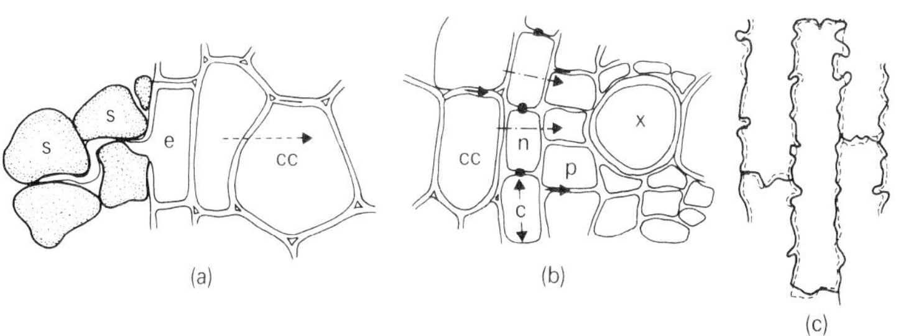

Fig. 4.5 *Left and centre* Semi-diagrammatic representation of the pathway of water and solutes entering a *Ricinus* root. In soil(s) water and ions move in unison but in the epidermis (e) and cortex (cc) ions may move from cell to cell via plasmodesmata, while solutions may move freely in the walls up to the endodermis (n) where Casparian bands prevent 'free space' access. Uncertainty remains concerning flow from the endodermis protoplast (n) via pericycle (p) to xylem conduits (x). *Right* Casparian band 'net' remaining after digestion of a whole root in concentrated sulphuric acid.

brick-shaped endodermal cells as an extensive meshwork of strips surrounding the radial walls of the endodermal cells, providing an effective cell-wall barrier of suberin impregnated with fats and other hydrophobic substances. At the endodermis both water and solutes must pass through cytoplasm which is so firmly connected with the Casparian strips that contact remains even in preserved or plasmolysed material.

It does not follow that ion uptake *must* occur only at the endodermis, indeed there is excellent evidence showing that protoplasts of cortical and epidermal cells including root hair cells absorb solutes from the aqueous solution passing via their walls. Once absorbed into the symplasm, transport from protoplast probably follows the plasmodesmata into the stele, until at some point the solutes are released into the xylem. The mechanism and location of solute secretion are still highly controversial. Evidence has tended to favour the role of cells such as xylem parenchyma, and not the endodermis, on account of the paucity of metabolic machinery, such as mitochondria there, which might be required for solute secretion. Another suggestion is that immature xylem initials might fulfil this function. However achieved, solutes are eventually secreted into the xylem, the individual ionic composition of which is often many times that of the concentration of the external medium bathing the root.

It may be observed that in primary roots phloem and xylem tissues alternate. The importation of starch and other anabolites is undoubtedly effected from leaves via the sieve tubes. Flow in the phloem is agreed to occur by bulk flow of solution and import must be accompanied by water from source tissues (see Ch. 9). The critical role of the phloem to sustain ion uptake has been clearly demonstrated by bark-ringing the plant (Bowling, 1965).

The root as an osmometer

Since antiquity it has been observed that decapitated root stumps often exude water. If attached to a manometer, considerable pressure can be generated, and though this seldom exceeds 1–2 bar much larger pressures have been measured. For example Lyon observed a pressure of 2–3 bar in a leafless birch, a pressure more than sufficient to raise sap to the top of the tree. Similarly in experiments on *Ricinus* plants 1 m tall we have measured pressures directly up to 2.2 bar. White (1938) found tomato root tips were able to generate pressures up to 10 bar. Nevertheless attempts to explain the flow of sap by this process have not been successful. Sap pressures are invariably *negative* when transpiration rates are maximum and tend to rise slowly even in the absence of transpiration, because the flow rate through root

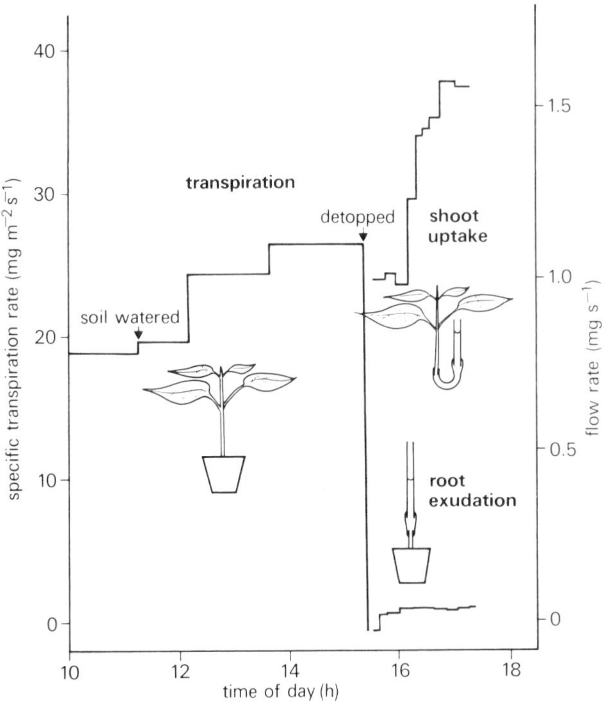

Fig. 4.6 Transpiration of a *Ricinus* plant in a growth chamber was determined by periodically weighing the whole pot. The plant was detopped and potometers fitted to the shoot and root. At first the root absorbed water, but even when it began to exude the rate of flow could not match the demands of the transpiring shoot. This experiment (first performed by Kramer, 1939) proves that root pressure cannot account for transpiration.

pressure is incapable of meeting the demands of a transpiring shoot (Fig. 4.6).

Another aspect of osmotic behaviour of roots is *guttation*, the process whereby plants exude water from leaves when transpiration becomes insignificant, usually at night. This process is often erroneously attributed to dew deposition from the atmosphere. Guttation fluid differs from root stump exudation because the leaf exudate has a much lower solute concentration. *Colocasia antiquorum* (the taro) was observed by Dixon (1914) to guttate at a remarkable rate of 10 ml per night! Solutes were not detected in the liquid collected; typically many solutes are extracted from the guttation stream during passage through stem and foliar tissues.

Argument persists as to the utility of such osmotic behaviour. Some suppose that osmotic pressures develop incidentally from the accumulation of solutes, the benefit of which is obvious. Alternatively the guttation stream could facilitate solute distribution, but the flow rate is insignificant in comparison with that during transpiration. A quite different suggestion, that compression of the xylem sap might gradually remove gas emboli, especially if gas is at very low pressure following cavitation, has received support from the observation that many small herbs, e.g. *Plantago* or *Tussilago* spp, which guttate freely also cavitate freely by day (Milburn and McLaughlin, 1974).

When the whole root system is considered as an osmometer, it can also be compared with a single cell. According to this analogy the cortex is compared with the cell wall, both of which are accessible to the free diffusion of water and solutes; the endodermal cylinder is equated with the semi-permeable membrane or the plasmalemma; the stele is equated with the vacuole.

Sabinin (1925) proposed that root exudation was proportional to the solute difference between the solute potential inside Ψ_{si} and outside Ψ_{so} a root system.

Exudation flux $J_v \alpha \Psi_{si} - \Psi_{so}$, or $J_v = k(\Psi_{si} - \Psi_{so})$ [4.2]

This simple relationship has been supported by many experiments, but before considering the root further as a simple osmometer it is necessary to examine evidence not easily reconcilable with this view.

Factors influencing J_v the hydraulic flux through roots

Diurnal rhythms

The rate of sap flow from detopped root systems often exhibits circadian rhythms as reported by Grossenbacher (1939). We have recently confirmed that this applies to exudation pressure of *Ricinus* root stumps in pots of soil. No osmometer should exhibit such fluctuations while temperatures and water potentials are constant so metabolic activity must be involved.

Inhibitors

Many inhibitors reduce the uptake of water, e.g. KCN. There is a known inhibition of salt uptake by tissue slices, so the implications are that as salt uptake is reduced water uptake is reduced accordingly. This evidence implies that the root is more than a simple osmometer. If water transport is affected in step with ion secretion, we may hypothesise that the osmometer is being affected through its capacity to regulate Ψ_{si} through transport linked to metabolism and respiration.

Temperature

It has long been known that if plants are watered with cold water during rapid transpiration they often wilt. Sachs (1882) showed that this effect was caused by low temperature. The fact that temperature reduces the flow is not surprising; we should expect some effect from the increase in viscosity which increases about 50 per cent from 20°–0°C. The puzzling aspect is that the *extent* of uptake is much greater

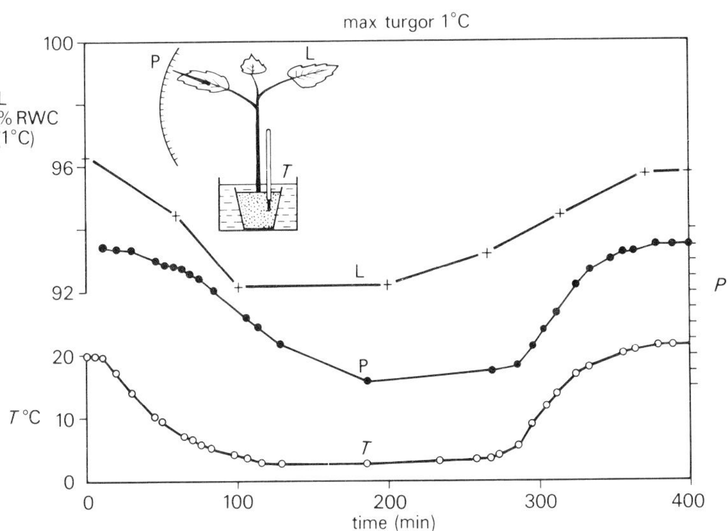

Fig. 4.7 An experiment on a transpiring *Ricinus* plant to investigate the effect of root cooling T on wilting, measured from the angular decline of the petioles P, and the water deficit of the leaves L determined by punching leaf discs and measuring their relative water content RWC. Discs were weighed before and after floating on water (4 hours) to full turgor at 1°C. This temperature was chosen to overcome growth expansion (from Milburn, 1964). *Inset:* Sketch of experimental system.

than 50 per cent in the case of tropical plants, e.g. *Ricinus*, nor is the effect linear; while temperate plants may be hardly affected. It may be observed that the effect makes a very convenient and reversible 'valve' by which root uptake of water may be regulated (Fig. 4.7). One possible explanation for the effect of temperature is that cytoplasmic processes in the root such as the hydraulic conductivity of endodermal membranes are influenced directly.

Table 4.2 Measured rates of hydraulic flux J_v into roots expressed in terms of uptake at the root *epidermal* (or root hair) surfaces. Water potential gradients $\Delta\Psi/x$ were not measured but were unlikely to exceed 25 bar. Note that root hair uptake does not necessitate protoplast uptake and that there can be considerable uptake by suberised roots (derived from Kramer 1969 and Anderson 1976). Hydraulic conductivity L has been calculated for the *endodermis* of each root assuming (1) that the endodermis is the critical (i.e. only) resistance to flow; (2) its thickness x is 10 μm; (3) that $\Delta\Psi = 10$ bar; and (4) that its surface area is one-fifth of the epidermal surface.

Plant material used	Conditions of experimental material	Hydraulic flux J_v of epidermis $\times 10^{-8}$ m^3 m^{-2} s^{-1}
Onion	Young roots in water	14.0
Onion	Roots in 1 mM KCl	36.0
Maize	Young roots in water	5.5
Maize	Roots in 1 mM KCl	72.2
Oat	Roots in 1 mM KCl	33.2
Radish	Root hairs in water	5.1
Sour orange	Suberised roots in water	1.3
Short leaf pine	Suberised roots in water	0.9
Coffee tree	Roots in soil	0.06
Various species roots	–	5.61

Respiration

Flooding root systems can induce wilting also, an effect now shown to be caused by anaerobiosis – a shortage of oxygen accompanied with a toxic build up of carbon dioxide (see Ch. 10). While it is reasonable to suppose that turgor might be influenced by a suspension of ion accumulation in a transpiring plant, water is absorbed in response to negative xylem sap pressure, so that $\Psi_p > \Psi_s$ in the root by a considerable margin. Clearly respiration is able to reduce water transport directly without implicating ionic effects. The effect must be attributable to living cytoplasm because it can be reversed by aeration. Kramer has shown that dead roots, killed by boiling, have a much greater hydraulic conductivity than when living.

Pressure

In transpiring plants water is drawn through roots in response to considerable pressure gradients produced by the transpiration stream, an effect difficult to simulate under laboratory conditions. Mees and Weatherley (1957), however, established measurable pressure gradients across a root system by enclosing it in a pressure chamber by pressurising the *outside* of the root above that of the exuding stump at atmospheric pressure.

Many curious results were found in these experiments. One surprising finding was that the hydraulic conductivity of a root system increased markedly in response to an increasing pressure gradient.

Hydraulic conductivity L of endodermis $\times 10^{-14}\,m^2\,bar^{-1}\,s^{-1}$	Authority and year
70	Rosene (1941)
180	Hay and Anderson (1972)
27.5	Hayward *et al.* (1942)
361	House and Findlay (1966)
166	Collins, J. C. (unpubl.)
–	Rosene (1943)
6.5	Hayward *et al.* (1942)
4.5	Kramer (1946)
0.3	Nutman (1934)
25–305	Newman (1976)

Root hydraulic conductivity

The topic of root permeability to water is complex and by no means fully explained. In an intact plant the flux of solutes cannot easily be separated from the flow of water. The flow of water in response to pressure differs from that produced by an osmotic gradient. Furthermore, roots can become more conductive in response to an increasing hydrostatic pressure gradient (see above). Results are difficult to express in terms of surface areas of the absorbing zones because there is no certainty that the most critical area resides at the endodermis though this seems most likely. Accordingly results have been expressed, for example, in terms of hydraulic conductance, L_p, across entire root systems or per length of root. Representative results given in Table 4.2 should be regarded as *minimum* values because they were measured using pressure gradients, $\Delta\Psi_p$, less than 1 bar, while transpiration would normally induce pressure gradients of 1–7 bar across root systems *in vivo*. Assumptions have been made to indicate the order of magnitude of the hydraulic conductivity when the critical resistance is assumed to lie near the endodermis. Alternatives to the endodermal barrier lie within the endodermal cylinder, e.g. the maturing xylem conduits.

Electrical gradients

In certain respects the whole root system can be regarded as if it were a single cell, vacuolar sap corresponding with the sap exuding from a decapitated root system. Studies of the electrical system have been made, using the Nernst equation (see Ch. 3) in relation to the uptake of ions (Bowling, Macklon and Spanswick, 1966). Electrical potentials in *Ricinus* roots are commonly −50 mV inside the root relative to the external bathing solution. It seems this may be achieved by proton extrusion pumps.

Recently this work has been extended using ion selective electrodes to study the uptake and transport across the cortical cells. There are gradients which suggest potassium uptake takes place across the whole of the cortex (Bowling, 1973), but others (e.g. Anderson and Higginbotham, 1975) have expressed doubt regarding the validity of these measurements. There is no unambiguous evidence that the electrical gradients might propel water other than indirectly, i.e. by osmosis. The important point at issue is the location of ion and water barriers. Logically they might be expected to have the same location but experimental evidence is conflicting and may indicate that our concepts need modification.

Chapter 5

Xylem – the vulnerable pipeline

Xylem is the name given to a plant tissue, the wood, which functions as the main water pipeline in plants. Xylem is composed of many cell types of differing structure and function. Long distance water transport occurs mainly in recently formed xylem called sapwood. In perennials it becomes clogged by air, tyloses and complex polymers after a number of years, becoming heartwood, when it ceases to conduct significant amounts of water. Nevertheless, both heartwood and sapwood are significant structurally.

Ray cells, also part of the xylem tissue, allow radial transport and also function as living stores of metabolites. Starch may be stored in the xylem to the extent that it is worth extracting commercially, e.g. from *Metroxylon sagu* the sago palm. Other storage products are released into the sap, particularly at certain times of the year, for example sugar in the sugar maple (*Acer saccharum*) in spring. This phenomenon of solute exchange via the transpiration stream is probably more widespread and important than generally realised, nor must it be forgotten that all cells living (parenchyma and secondary xylem initial cells, etc.) and dead (fibres, tracheids and vessels) function as a water storage system. The compensating action of this store on the functioning of the whole plant may be very important in smoothing out rapid fluctuations in sap pressure. In the short term this may protect the plant from cavitation, which interrupts sap conduction by the development of gas emboli in the conduits, and in the longer term allow time for the metabolism to adjust to a new water regime.

It is interesting to see how sap transport, appreciated as a problem since ancient times, has been resolved with the passage of time. Inseparable from the problem of sap conduction itself is the nature of the matrix through which it passes. Many otherwise competent investigations have foundered owing to ignorance of the anatomy of conducting tissue.

Historical development

The water requirement of plants has been understood since the dawn of civilisation. Doubtless those who irrigated the Hanging Gardens of Babylon by raising water 25 m from the Euphrates had ample opportunity to reflect on this point! Up to the twentieth century the problem was mainly conceived as one of raising considerable volumes of water through tree trunks – hence the *ascent* of sap. More recently the overriding importance of soil moisture status has become recognised, particularly in saline soils (e.g. for the mangrove) and in deserts. Here the problem is more a matter of water extraction rather than elevation against the relatively weak gravitational forces.

Overall, the historical picture is one in which theories of physical mechanism have gradually gained ascendancy over vitalistic theories (pumping by living cells). Physical processes in plants have often differed from those commonly studied in physics courses and many observations, e.g. cohesion of sap, seem at first to contravene the established laws. The Chronological Synopsis indicates to an extent the way discoveries have interacted. Certain experiments have been repeated, frequently without recognition of earlier work, e.g. the overlapping saw cut experiments, the ascent of poisons, and double-shear sampling.

It is emphasised that the process is now understood to be almost wholly physical, but the establishment of water-filled conduits from meristematic initials is of course dependent on living cells. Living cells can protrude tyloses into adjoining conduits to prevent conduction. We are not yet sure of the extent to which living cells intervene to protect water columns from damage by remote action, e.g. by releasing solutes or absorbing gases to enhance frost resistance of sap.

Structure of the conducting pathway

Three xylem cell types conduct water, vessel elements, tracheids and fibres. They conduct when only the dead cell wall remains. A *tracheid* consists of a spindle-shaped tube which is rather short in length (commonly 3–5 mm) and originates from a single cell. *Fibres* are similar; they have characteristically smaller lumina, fewer pits and thicker walls than tracheids. The distinction between fibres and tracheids is somewhat arbitrary. A *vessel* consists of the cell walls of vessel elements cemented end-to-end to form a hollow tube. It may be short, or very long (Table 5.1). Vessels and tracheids normally end obliquely with walls richly endowed with pits through which sap (but not gas or particles) can pass into a similar overlapping end of the next conducting unit. It is convenient to describe *all* of these conducting units (fibres, tracheids and vessels) by the general term

Table 5.1 Maximum vessel lengths. Various hardwoods of Nova Scotia (after Greenidge, 1952).

Wood type	Species	Maximum vessel length (m)	Tree height (m)
Ring porous	*Quercus borealis*	15.24	15.85
Ring porous	*Fraxinus americana*	18.29	19.05
Ring porous	*Ulmus americana*	8.53	9.14
Diffuse porous	*Acer saccharum*	0.94	–
Diffuse porous	*Acer rubrum*	0.95	–
Diffuse porous	*Betula lutea*	1.42	–
Diffuse porous	*Betula papyrifera*	1.04	–
Diffuse porous	*Fagus grandifolia*	5.56	–
Diffuse porous	*Populus tremuloides*	1.32	–
Diffuse porous	*Prunus pennsylvanica*	1.45	–
Diffuse porous	*Alnus rugosa*	1.22	–

conduit. Physiologically a conduit functions as a water-filled spindle separated from other conduits by a membrane system.

As conduits develop from meristematic initials their water balance passes through two main phases. Before walls are lignified the thin-walled cell expands the living protoplast developing a positive turgor pressure by water absorption. Lignin is then deposited, while cytoplasm and living membranes are largely digested and disappear (see Fig. 5.1). As this happens the water-filled conduit becomes integrated in the conducting system with other conduits and the internal pressure tends to become negative. Submerged aquatics are exceptional because the internal sap pressure is positive even when mature; lignin is not deposited and xylem vessels have walls resembling those of sieve tubes.

Conduits do not seem to bifurcate in higher plants. However, the vascular traces themselves often do so. An interesting observation by Zimmermann and Tomlinson (1967) is that conduits are not nearly so parallel as generally assumed. The structure has been beautifully demonstrated by cinemicrography of the vessels. Sections are cut along a length of wood. Carefully aligned micrographs, taken after each section is removed, are run through a cine-projector, allowing

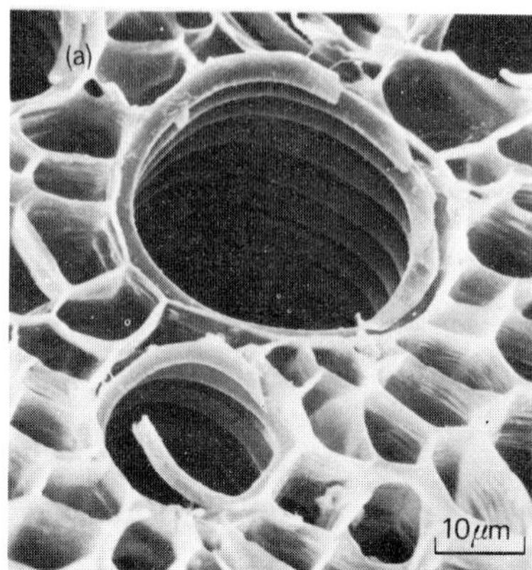

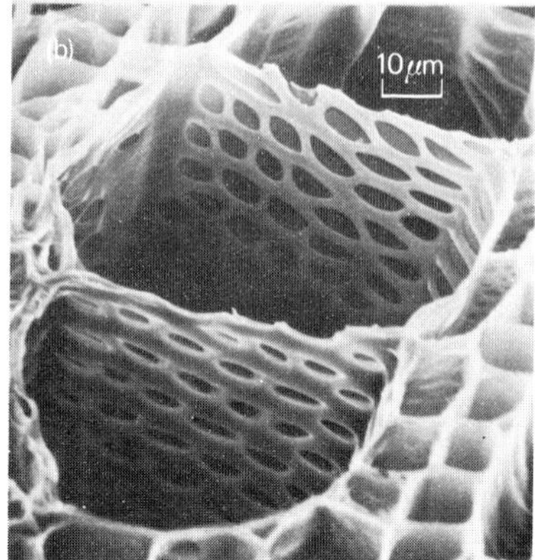

Fig. 5.1 Xylem vessels of *Ricinus* petiole showing (*left*) how the conduits are strengthened by spiral thickening in a petiole to permit extension growth. When vessels are laid down in mature tissues (*right*) the thickening is scalariform (ladderlike) or reticulate. The large bordered pits are closed by membranes which allow lateral as well as longitudinal sap flow (Scanning electron microscopy [Milburn and Sprent, unpublished]).

three-dimensional exploration and analysis of the xylem. Individual vessels can be seen to pass over their neighbours to an extent which depends on the species, being very marked in *Fraxinus*. The wood grain may deviate also from the vertical in different directions in successive years producing a crossover grain, described as 'interlocked'.

Role of pits

Pits are of two main types, simple and bordered. In xylem conducting tissue the bordered *pit-pair*, consisting of two aligned pits from adjacent cell walls, predominates. In general, pits act so as to protect the conducting system by their valve-like action. But why are they so complex?

Water is able to pass relatively freely through pit membranes, but most membranes filter out particles. However, the most important role is the prevention of the spread of vapour from one conducting unit to the next when a vessel cavitates or is punctured. When this happens the stress placed on each pit membrane is enormous – being approximately zero pressure inside with tensions up to 100 bar in adjoining vessels. If the wall were thick enough to withstand collapse its resistance to water flow would be very high. Apparently by having a thin and porous membrane (pores 0.1 μm radius; Eicke, 1954) with two borders, the membrane is strengthened without the resistance to normal flow being significantly increased. *Ricinus* and other angiosperms have simple pit membranes, but those of gymnosperms have a central *torus* which functions as a plug. Under stress the central torus is pressed against a border forming a seal when the pit is said to be *aspirated.*

Not all pits accord with this scheme. Indian ink, consisting of a fine suspension of carbon particles which are stopped by most pits, can traverse others (e.g. *Abies*) in which the pores are relatively large (see cover). Nevertheless gases are stopped by the surface tension acting on the membrane so that the torus seals the pit against a border. The way in which pit membranes function under different conditions with respect to gas flow has led to many misunderstandings in the past. Though wet pit membranes stop the spread of gases, dry membranes are freely permeable to gases.

An interface, e.g. gas–water, mercury–water or fat–water, is stopped by a porous hydrophilic membrane because of surface tension forces of water which act across each minute pore. (The size of the pores is smaller than indian-ink particles.) In the same way pits can become aspirated through clogging by dust particles in water passed through them under experimental conditions. Figure 5.2 shows results obtained when untreated water is forced through a length of petiole under three pressure regimes. The increasing resistance seems due to

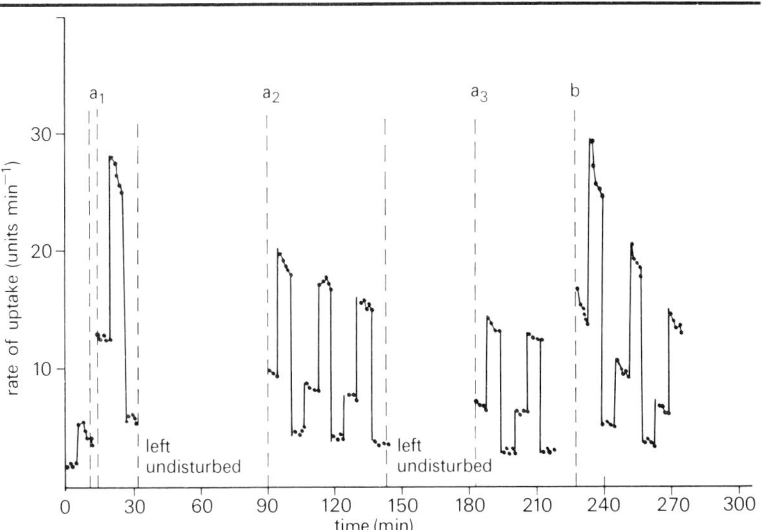

Fig. 5.2 Experiment investigating the flow of water through a detached *Pelagonium* petiole under three alternating suction pressures (200, 100 and 50 mm Hg). Water conduction decreased a_1, a_2, a_3 in proportion to the volume of water transmitted but also with time but could be restored ephemerally b, by cutting crushed end from the petiole but thereafter conduction decreased even more rapidly.

the quantity of water transmitted, hence the quantity of material filtered out, rather than the time which might be required for callus or tylose formation which could also block the conduits.

Florists have developed special techniques, such as crushing the ends of flower stalks, which expose more extensive areas of xylem for water absorption. Various additives to the water probably minimise the build-up of bacteria which tend to block the pores mechanically. Yet it is common practice to cut flowers in air when large air embolisms may invade the tracheary system. Some flowers do not survive, but how do *any* survive this disruption? Apparently there are usually sufficient undamaged conduits, still water-filled, adjacent to the cut stump to permit uptake. The surface tension of water acts on the bubbles now trapped in the damaged conduits. This increases their internal gas pressure which induces them to dissolve in the water restoring their ability to function. Clearly the shorter the average conduit the greater is the chance of water-filled conduits being adjacent to a cut. Thus shorter conduits are more reliable in such adverse conditions in contrast with the seemingly more efficient wider and longer conduits which are extremely vulnerable.

If water is added to the top of a piece of water-saturated stem, held

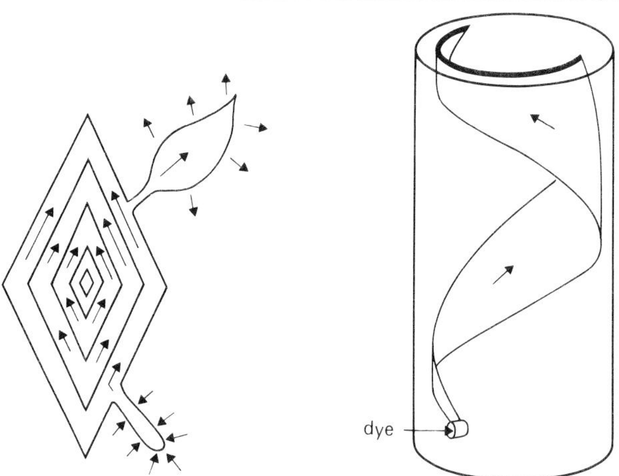

Fig. 5.3 Diagram to illustrate why water conduction tends to occur predominantly in the most recently formed conduits (*left*). Layers of conduits form shells around the older wood. When cut across, these show annual rings. In secondary thickening tissues, current leaves and roots are connected by the current year's wood. If liquids are introduced at pressures exceeding that of the sap they tend to spread tangentially and radially following the grain of the wood. Thus a piece of trunk with a spiral grain when injected below illustrates a spirally spread pattern as often found (*right diagram*).

vertically, a drop appears at the lower end. The effect can be enhanced by cutting sections from the uppermost end of the stem. This interesting observation was first explained by Godlewski (1884). After each cut water runs through the stem section until it hangs by surface tension on air–water menisci in the undamaged conduit wall's pit membranes, which prevent further expulsion of water by air.

Pathway of water movement

In roots the xylem pathway is diffuse, becoming progressively aggregated in the stem. After passing along the pipelines of the stem, the xylem tissues ramify once more in leaves among the mesophyll cells.

Generally speaking, in intact perennials most sap is transported in the outermost sapwood. This is to be expected since water enters the outermost wood in the roots, takes the path of least resistance up the stem and passes into the leaves through recently formed xylem (see Fig. 5.3). It applies particularly to ring-porous angiosperms but less to diffuse-porous angiosperms and still less to trees without vessels, the gymnosperms. When the conduits are short, pits occur frequently in a longitudinal direction and lateral flow is increased, the inner layers of wood becoming more important in normal conduction. Thus when dyes are injected into conifers the pattern of upward distribution is not necessarily peripheral (see Fig. 5.3). Nor is it necessarily vertical; it may spiral and even penetrate sectors tangentially depending on the orientation of the conduits and their resistance to water flow. Such patterns have been demonstrated in many trees by Greenidge and others. It is important to recognise that when liquid dyes are injected at atmospheric pressure ($\Psi_p = 0$ bar), since the xylem sap is under tension (say, $\Psi_p = -10$ bar), the water balance is disturbed. This causes dye to move upwards and downwards and also to spread laterally (Fig. 5.4) giving an impression that sap movement is diffuse – these effects are artificially produced.

In damaged stems, where the vessels are punctured or become blocked by tyloses or gas, the pathway of water movement changes. Thus sap passes round overlapping saw cuts as demonstrated by Hales (1727). 'Here again we see the very free lateral passage of sap, where the direct passage is several times intercepted.' It has been argued falsely, on the basis of over simplified models, that this treatment should completely eliminate the whole conducting system. Pfeffer (1897) explained this phenomenon clearly and correctly, ascribing the continued passage of water to the considerable numbers of short conduits remaining intact and water-filled between the cuts.

Water conduction through vessel walls

Sachs (1887) believed that all transpired water could be transported

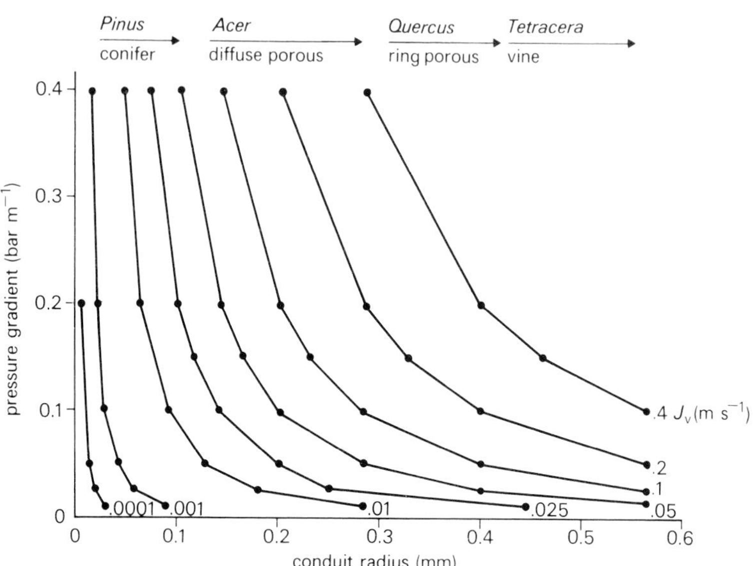

Fig. 5.4 Relationship between pressure gradients, and sap velocity J_v in different conduits of different radii on the assumption that the conduits are endless capillary tubes and amenable to Poiseuille analysis (adapted from Zimmermann, 1964).

through vessel walls in the imbibed condition. Simple experiments, in which the lumina were blocked, showed that wall conduction was much too slow to permit the observed flow. Nevertheless lignified walls are porous and the pores are usually water-filled (up to 50 per cent of the dry weight). The structure of the attracted water changes due to adhesion between water molecules and the wall matrix reducing its freezing point below that of free water 0°C. Thus even when transport through conduit lumina is impossible significant quantities of water may yet pass through the vessel walls, explaining the observation by Hygen (1963) that frozen woods still conduct dyes below 0°C.

Mechanism of sap transport

There has been a gradual shift from vitalistic towards physical mechanisms for the explanation of sap transport. The Askenasy demonstration which shows how evaporational forces may raise water strongly, so that water and mercury columns can be raised to considerable heights, provides an adequate model (see Ch. 2).

In terms of water potential one may say that water moves along a gradient of potential from soil to the atmosphere. The normal situation can be represented in this continuum:

ΨPure water→ΨSoil→ΨRoot→ΨXylem→ΨLeaf→ΨAtmosphere

Decreasing water potential

————————————————————→

Increasing water stress

Root pressure, which so fascinated earlier workers, is now recognised as a relatively unimportant mechanism of sap transport (see Ch. 4). The flow through the root is insufficient to meet the demands of a transpiring shoot. The most likely benefit of root pressure is its role in refilling cavitated conduits in seedlings and herbaceous plants.

The cohesion theory

This theory, developed more or less independently by Dixon and Joly (1894) was supported strongly by Askenasy's (1895) demonstration of the capability of physical forces to elevate columns of water. In modern form the concepts are as follows:

1. Liquid water within a whole intact plant is regarded as a continuous single phase extending from the microcapillaries of the mesophyll cell walls via the xylem conducting system to the microcapillaries in cell walls in the roots.
2. The bore of cell wall microcapillaries is sufficiently narrow to enable the surface tension of water at the menisci to resist the entry of atmospheric air yet permit water to evaporate from them by transpiration. The free energy of water equivalent to the intramolecular pressure is reduced depending on the meniscus curvature; free energy is maximum if a meniscus is flat, but reduced if negative pressure induces curvature.
3. Columns of water persist in xylem vessels and tracheids owing to the cohesion between water molecules and adhesion with the walls of the tubes.
4. Energy for movement of the water catena is provided by evaporation from the leaves, providing that the free energy of water in the leaf microcapillaries exceeds that of water vapour in the surrounding atmosphere.
5. The cohesion theory can be extended beyond the plant into the soil matrix owing to the excellent contact between root hairs and the surface water films surrounding soil particles.

Role of living cells

Curiously, Dixon, who played such a vigorous part in the establishment of the cohesion theory, believed in the vital agency of mesophyll cells (see Ch. 8) but there is now no reason to believe that living cells participate directly in the sap transport mechanism.

Chronological Synopsis up to 1965

*, ° *and* † *indicate repetitions of same type of experiment.*

1628	*Harvey*	*Propounded circulations of blood.*
1660	*Malpighi*	*Discovered capillaries confirming circulation of blood in animals.*
1669	*Ray*	*Sap circulates in plants like blood in animals.*
1709	*Magnol*	*Sap pathway shown by dyes.*
*1717**	*Hales*	*Quantitative approach to sap transport in plants (and animals).*
*1765**	*Duhamel*	*Sap ascent continues despite overlapping cuts.*
1840°	*Boucherie*	*Sap ascent continues despite poisons.*
1882†	*Volkens*	*True sap content of vessels by simultaneous double cut shears.*
1884	*Godlewski*	*Cells must pump water to account for sap movement.*
1887	*Sachs*	*Vessels empty; therefore water moves physically through imbibed walls.*
1891°*	*Strasburger*	*Sap ascent despite (1) poisons then dyes (2) overlapping cuts.*
1894	*Dixon and Joly*	*Cohesion theory propounded for ascent of sap.*
1895	*Askenasy*	*Improved Detmer apparatus to show considerable forces of cohesion.*
1914	*Dixon*	*Transpirational function and gravitational pull affect sap ascent equally.*
1925	*McDougal*	*Diurnal contraction of tree trunks by denography.*
1927	*Bose*	*Sap ascent by pulsation of cortical cells.*
1935	*Priestly*	*Doubted cohesion mechanism: xylem air-filled.*
*1936**	*Elazari-Volcani*	*Overlapping cuts do not prevent sap ascent.*
1937	*Huber and Schmidt*	*Velocity of sap flux by heat pulse method.*
1938	*Preston*	*Estimation of sap tensions by rate of ink injection.*

1938	*Dixon and Barlee*	*Leaves exude water by vital action of leaf cells.*
1939	*Crafts*	*Water columns* (*in* Ribes) *cavitate if jarred.*
1939	*Handley*	*Cooling of stems above freezing point causes wilting.*
1942†	*McDermott*	*True xylem sap content of wood by simultaneous cuts.*
*1950**	*Preston*	*Overlapping cuts. Trees did not always die.*
*1955**	*Greenidge*	*Overlapping cuts. Dyes moved round cuts.*
*1958**	*Postlethwaite and Rogers*	*Overlapping cuts.* $^{32}PO_4$ *moved round cuts.*
1964	*Zimmermann*	*Handley's experiments disproved by improved techniques.*
1965	*Scholander* et al.	*Measurement of negative sap pressures by pressure bomb.*

Conduit length, girth and hydraulic conductivity

Under normal conditions vessels are undoubtedly the main water conducting conduits in land plants. At the end of a conduit water must pass through pit membranes which are essentially the remnants of the primary walls. Pit membranes are finely porous causing resistance to water flow but benefit the plant by limiting cavitation and also delay invasion by pathogens. The study of conduit length is important for these reasons. Maximum conduit length is not particularly relevant in most studies, what is needed is a population study. Exactly the same principles apply to conduit girth because the resistance to longitudinal flow is strongly dependent on the radius of the conduits.

Several methods can be used to measure individual maximum conduit lengths based on the transmission of mercury, gas or indian-ink (a suspension) through the conduits. Population data have been determined using paint or ink injection. A cut stalk is injected with ink then sections cut from the stump end and the percentage of ink-filled conduits is determined by microscopic examination of transverse sections cut at regular intervals. The results are plotted as per cent injected, against stem length injected forming an 'injection profile'. The interpretation of this profile has been described by Milburn and Covey-Crump (1971) and used successfully by Mackay and

Weatherley (1973). The method is easy to apply; its precision is limited however, and a more precise technique may be desirable. The most precise method utilises the 'shuttle' microscope (Zimmermann and Tomlinson, 1965). A single ocular views two separate serial sections which are aligned, while switching their source of illumination, then photographed sequentially to give a cine-film. This film is projected using a projector which can be stopped and reversed allowing an extremely comprehensive, if tedious, analysis. A simpler system also utilises a cine-camera but the surface ends of stems or petioles are photographed each time a new section has been cut from them. The latter method requires specially rigid support and cutting and illumination systems. As yet these methods have been used to study vascular pathways rather than more detailed studies of conduit populations (Tomlinson and Zimmermann, 1967, Zimmermann, 1976).

The hydraulic conductivity of wood depends on the bore number of the conduits, and the frequency of conduit endings within them. The radii and number of conduits can be measured easily from transverse sections. Flow through the conduits has been analysed using the Hagen–Poiseuille equation according to which the volume of water flow in each conduit is dependent on the fourth power of the radius. From the equation L and L_p can be calculated

$$J=\frac{V}{t} \quad (=J_v A) \quad =\frac{\Pi r^4}{8}\cdot\frac{\Delta P}{x} \quad \left(=LA\frac{\Delta P}{x}=L_p A\Delta P\right) \qquad [5.1]$$

(in the usual units, see Ch. 1 and App. 1).

Such is the effect of the term r^4 that a small proportion of large bore conduits carry the bulk of flow in a mixed population of conduits (Fig. 5.4), an effect probably accentuated by the fact that in wide conduits there are fewer pit membranes to pass per unit length. A Hagen–Poiseuille analysis, of the type applied by Dimond (1965) to study sap flow in tomato stems, illustrates this point clearly for *Ricinus* stems (Fig. 5.5).

Hydraulic conductivity of woody stems can also be determined directly by experiment. Low pressure, ΔP, drives water flow, J_v, through a specimen of wood (length x and cross-section A) and the hydraulic conductivity, L, is calculated from the relationship in the standard units (App. 1).

$$J_v=\frac{V}{At}=\frac{L\Delta P}{x}, \text{ so that } \frac{V}{t}=LA\Delta P \quad (=J_v A=J) \qquad [5.2]$$

Typical values (see Zimmermann and Brown, 1971) reflect expectations from 10 to 100 per cent of Hagen–Poiseuille analysis to stem flow; thus the hydraulic conductivity of conifers < diffuse-porous trees < ring-porous trees < vines (see Fig. 5.4, p. 88).

Table 5.2 Hydraulic conductivity L of wood of different plants, i.e. volume flow of water per pressure difference per unit length per cross-sectional area. The apparent hydraulic conductivity of the xylem lumina is calculated assuming that one-fifth of the xylem transverse section consists of conducting conduits. (Note: $10^{-4}\,m^2\,bar^{-1}\,s^1 = 1\,cm^2\,bar^{-1}\,s^{-1}$.)

Wood type or common name	Specific names or details of condition	Hydraulic conductivity L of wood (whole xylem) $10^{-4} \times m^2\,bar^{-1}\,s^{-1}$	Hydraulic conductivity L of xylem lumina $10^{-4} \times m^2\,bar^{-1}\,s^{-1}$	Author and year
Conifers	Softwoods	0.6	2.7	(See Huber, 1956)
Deciduous	Winterwood	0.4–1.3	1.9–6.6	(See Huber, 1956)
Deciduous	Summerwood	1.8–3.5	8.9–17.5	(See Huber, 1956)
Lianas	Vines	6.5–34.9	32.4–174.5	(See Huber, 1956)
Names				
Yew	*Taxus baccata*	2.0	10.01	(See Dixon, 1914)
Poplar	*Populus nigra*	9.5	47.5	(See Heine, 1970)
Willow	*Salix atrocinera*	1.3–1.8	6.5–9.0	(See Peel, 1965)
Ash	*Fraxinus excelsior*	5.9–6.7	29.5–33.5	(See Peel, 1965)

Undoubtedly the values in Table 5.2 reflect the approximate hydraulic conductivity of intact stems, but there are many pitfalls in making such measurements. In an excised branch the pressure gradient through both inner and outer layers of xylem is made identical, whereas in the intact plant conduction would be weighted in favour of the outermost layers because of radial resistances to flow (see Fig. 5.3). Even when clean water is passed through a specimen it tends to induce blockage from suspended dust particles (see Fig. 5.2), and if the applied pressure gradient is abnormally great, conduction especially in conifers may be reduced by aspiration of the tori in their bordered pits. The specimen length is important because a specimen must be longer than the length of the longest vessel. Otherwise vessels which have had the terminal pit membranes excised can carry the bulk transport in a manner analogous to an electrical 'short'. The significance of conduit length is discussed in Chapter 9.

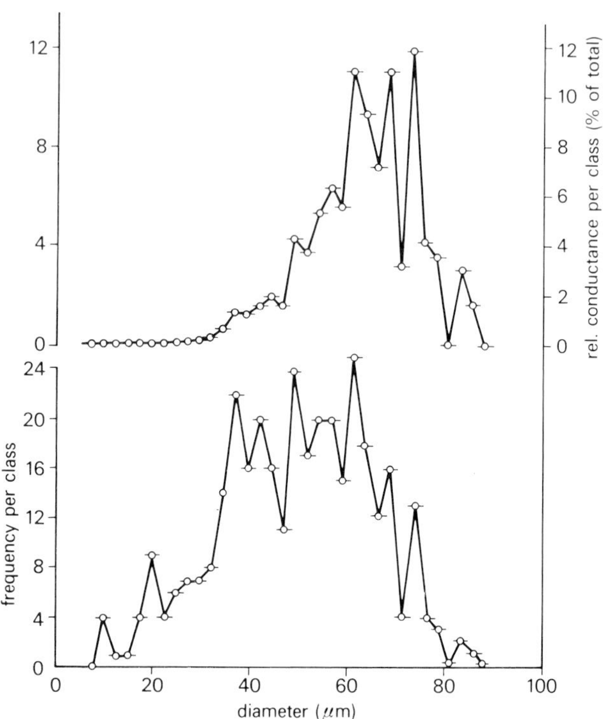

Fig. 5.5 Frequency of conduits of different diameter of a *Ricinus* petiole (lower curve). When analysed using Poiseuille equation the prediction is that most conduction occurs through a relatively small number of wider conduits (Milburn and Perrie, unpublished).

Measurement of sap flow J_v

A great many methods have been applied to measure sap flow in plants. Many simple techniques necessitate surgery; for example, when measuring the rate of transport using a suitable due as tracer (see previous section) or radioisotope (such as ^{32}P) or by attaching a potometer to measure the uptake of water directly. The techniques suffer from the fact that sap flow in damaged plants can be very different from that in intact plants. A more legitimate approach is to measure the uptake of water from a lysimeter or to determine transpiration from the whole plant gravimetrically. Though both of these methods are in use to study sap flow in large trees the equipment is cumbersome and expensive. Furthermore it is only possible to measure a flow rate averaged over lengthy periods of time when brief variations in the rate of sap flow may be under investigation; nor can such comprehensive methods be used to measure variations in sap flow in different parts of a tree.

To provide greater accuracy a number of methods have been devised to permit measurements on intact essentially undisturbed plants. (Similar methods are being employed medically to measure blood flow where it is also undesirable to wound the subject.) Of these the heat pulse method has been most widely used (see Fig. 5.6). In its simplest form a heater warms a small pulse of sap and this pulse is detected as it passes an electronic thermistor or thermocouple a measured distance away. Complications arise from the fact that the heat pulse must penetrate into and return from the stream being measured. Also the leading edge of the pulse becomes eroded through

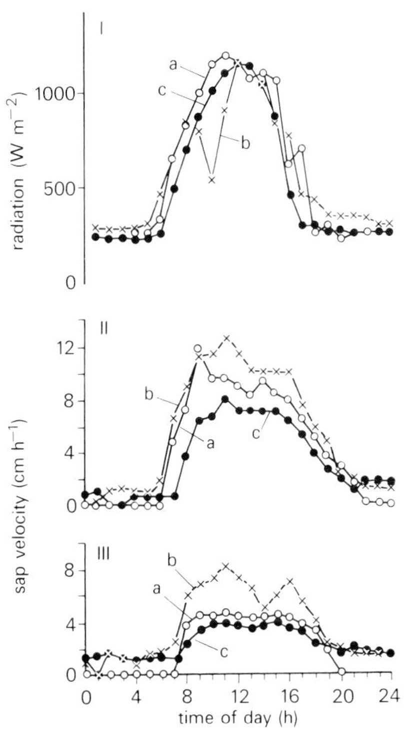

Fig. 5.6 Plot of sap velocity J_v, from heat pulse detection, in response to solar radiation (I); sap velocity in two species Englemann spruce (II); sap velocity of Lodgepole pine (III), on three dates in months (*a*) June (*b*) July and (*c*) September of 1964 (from Swanson, 1965). The correlation between solar radiation and sap velocity is obvious. Note the peak velocity of 12 cm hr^{-1} equals 0.03×10^{-3} m s^{-1}.

warming the surrounding tissues. In addition sap fluxes differ, even in adjacent vessels, making the heat pulse harder to detect. Heat pulse detection systems are either used to measure relative flow or are calibrated against specimens in which a measured water flux is induced artificially. The method is very successful for higher flux rates, but slow fluxes are difficult to distinguish from the heat flux of the pulse itself (Table 5.3).

Table 5.3 Midday mean velocities ($\bar{u}=J_v$) in trees, adapted from Zimmermann and Brown 1974, determined by Huber and Schmidt using heat pulse detection method. The values indicate the range from many different estimations.

Conduction system	Typical genera	Peak mean velocity $=J_v \times 10^{-3}\,m\,s^{-1}$
Ring porous spp.	*Fraxinus, Ulmus*	1.1–12.1
Diffuse porous	*Populus, Acer, Fagus*	0.2–1.7
Conifer	*Pinus, Picea, Larix*	0.3–0.6

A recent method showing considerable promise devised by Tyree and Zimmermann (1971), is based on the fact that a minute electrical current is generated by the flow of sap. It is called the Delta I (ΔI) technique. Another recent method is the magneto hydrodynamic flow meter, used to study blood flow for some time, which has now been successfully applied to plants by Sherriff (1972). This method is based on the fact that when a stream of water moves between the poles of an electromagnet a voltage is detectable in a secondary coil, as in a dynamo. The voltage is small, as might be expected from the small dipole of water molecules, and careful screening is necessary to avoid interference. The hydrodynamic method is very suitable for work on small plants, but as yet it has proved difficult to apply to forest trees. Both the above techniques give surprisingly linear curves when calibrated against known fluxes.

One of the most important reasons for measuring water flow in stands of vegetation is to assess the consumption of water in production of crops and timber. The cost efficiency of different stands in relation to water consumption is a key issue in studies on the cost effectiveness of irrigation.

Measurement of sap pressures Ψ_p and pressure gradients $\Delta\Psi_p x^{-1}$

Positive xylem pressures occur in trees relatively infrequently, usually in deciduous trees (e.g. maples or birches) early in the growing season. These pressures can be measured relatively easily using bourdon-type or mercury-manometer gauges. The main source of inaccuracy with pressure gauges depends on the volume of liquid to produce a true reading – gauges read accurately when this volume is small in relation to the volume and elasticity of the reservoir.

Pressure bomb

Negative pressure measurements are virtually impossible using standard pressure gauges, because any minute gas spaces in the gauge break the liquid continuity, so that negative pressures cannot be measured beyond vacuum. The problem was partially solved by Dixon (1914) but not until Scholander *et al.* (1965) produced their pressure bomb was a reliable method established. Pressure bombs are now indispensable equipment in studies of plant water relationships (Fig. 5.7).

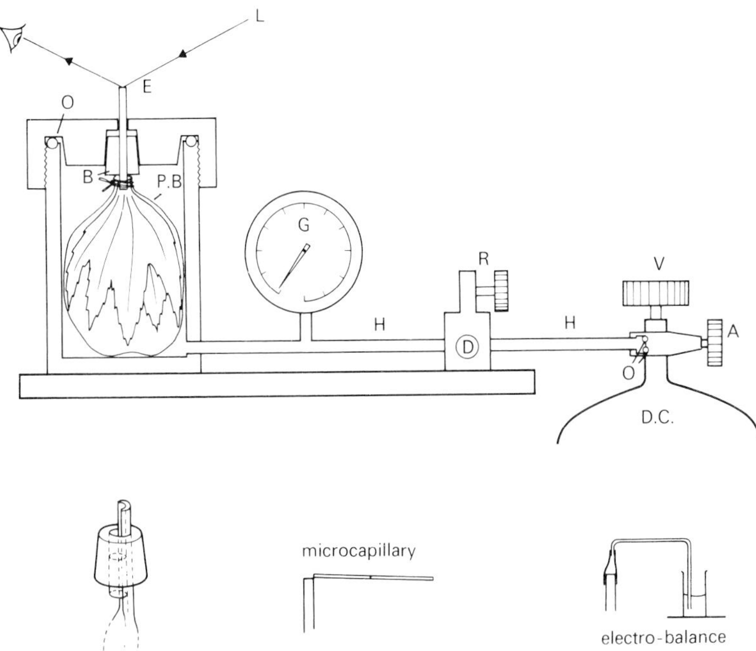

Fig. 5.7 Pressure bomb used to determine xylem tensions in a *Ricinus* leaf. The bomb lid is sealed by O rings and the petiole is sealed into the lid. Exudation E is viewed from the side by light L reflection. Compressed air from the portable diving cylinder D.C. connected by A clamp is fed by valve V through flexible hose H. Pressures are read on the pressure gauge G and the leaf is still enclosed in a plastic bag PB. *Inset below:* Double bung seal used for flattened petioles (*left*), exudate collection by microcapillary (*centre*) and by electro-balance (*right*). A bursting disc assembly, required for safe useage, is not shown.

When a shoot is cut from a plant the sap tension rapidly draws air into the conduits as sap retreats from the wounded surface until it is arrested by a conduit ending. If the shoot is sealed in a pressure bomb with the excised end projecting and pressurised gas is applied to the shoot, at a certain point sap will be forced back to the excised surface. If the pressure is increased exudation then occurs (see Ch. 3). What is happening is that the living cells which have absorbed water from the xylem on excision are squeezed until the xylem is refilled. When sap exudation is detectable from the cut stump the positive pressure on the cells is recorded; it exactly balances the negative pressure which held the water in the sap-filled intact conduits. Often the need to refer to negative pressures is removed by describing sap pressures as xylem tensions.

There are many potential errors in performing pressure bomb measurement of xylem tension. A standard precaution observed for precise work is to prevent transpiration for one or more hours before excising the specimen. This is necessary because the mesophyll cells beyond the xylem have a lower Ψ_p than the xylem sap itself during transpiration. When transpiration is stopped by enclosing a leaf or shoot in a plastic bag all cells develop the same water potential as Ψ_p of the xylem.

Other problems arise: if the xylem had suffered from extensive cavitation before sampling, abnormally high pressures might be required to refill the conduits. If stomatal closure reduces the penetration of gas into leaves, the leaves should be punctured otherwise the whole leaf is mechanically crushed by pressure often producing higher exudation pressure than required for compression of mesophyll cells. It has been pointed out that gas compression can give rise to adiabatic heating within the bomb (see effects of temperature Ch. 6) and gas flow can produce abnormal evaporation. If tissues are crushed in making a pressure bomb seal, sap can be expelled directly. Gas channels in the specimen can give poor measurements through frothing of the expelled sap. It is a wise precaution when determining incipient exudation to raise the pressure beyond the balancing point to ensure that sap does in fact exude.

Despite occasional difficulties in obtaining reliable measurements the pressure bomb is a remarkably useful tool, with an accuracy often approaching $\pm$ 0.1 bar, and considerable quantities of data have been accumulated in the last decade. Sap tensions in trees have been shown to increase by day in step with transpiration, reaching peak values of the order of 15–25 bar. Somewhat surprisingly, herbs and crop plants in which the 'ascent' of sap is no problem, generate sap tensions of similar magnitude (see Milburn, 1974). In part the high sap fluxes in plants are responsible for high sap tensions because it is difficult to extract water at the required rate even from reasonably moist soils. In recent years, however, it has become apparent that the hydraulic conductance L_p of xylem itself in herbs and terminal branches in trees is not so high as had been supposed from isolated measurements on wood specimens and certain trees (see Helkvist *et al.*, 1974). At night xylem tensions fall but tend not to reach zero even in well watered plants. The Ψ_p readings when water is abundant are usually -2 to -4 bar. Further work determining pressure gradients in plants is needed before we can be sure of all the implications of these observations. It must, however, be remembered that determinations of gradients require many more bomb readings than are required for point determinations and presently whole groups of important plants, e.g. the palms, have suffered incredible neglect. (Our experiments have recently established *Cocos nucifera* sap tensions in the range 2 to 12 bar (Milburn and Zimmermann, 1977).)

Other devices are available for measurements of sap tensions, e.g. the 'Aquapot' soil tensiometer, which has been embedded in tree trunks, psychrometric and cryoscopic techniques. These have potential advantages in allowing continuous monitoring of xylem, but it seems they are unlikely to be used more extensively than the pressure bomb for some time.

Pressure gradients and vertical height

As originally conceived the main problem facing plants, especially tall trees, was the ascent of sap. Since the advent of new techniques, particularly the pressure bomb, accumulated evidence is now overwhelmingly in favour of the concept that negative pressure gradients cause sap ascent. It is easy to predict the negative pressure required to counteract gravity alone, since Ψ_p decreases by 1 bar for every 10.13 m ascent. To this must be added an additional pressure required to overcome the frictional resistance of the wood. The effect of friction would vary from zero when flow is zero to a maximum coinciding with the highest rates of transpiration. Dixon estimated this frictional component to equal that due to gravity so that the gradient would be twice the predicted gradient of about 0.1 bar m^{-1} elevation.

Evidence accumulated from pressure bomb studies supports the notion that negative gradients occur of the anticipated magnitude. Scholander *et al.* (1965) demonstrated the anticipated gradients by shooting branches from very tall trees (redwoods and firs) to heights of 82 m despite the inherent problems of the method, because branches lose water during their fall to ground level. The gradients accorded well with Scholander's positive pressure gradients, measured in grapevine before leaf emergence, being parallel with the theoretical gradient. Similar pressure gradients have also been measured at midday in leafless sugar maples (Fig. 5.8), discussed in Chapter 9, and

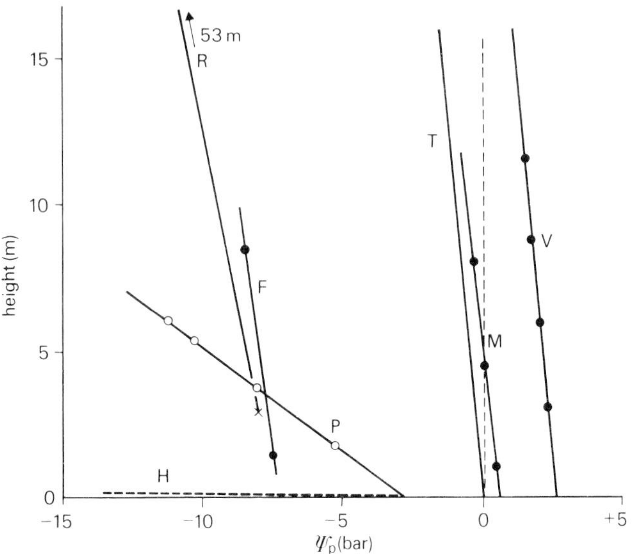

Fig. 5.8 Sap pressures in trees decrease with height (elevation). In grapevines V and leafless maples M at positive pressures the gradients are parallel with the theoretical gradient (10.13 m bar^{-1}). Redwoods R (*Sequoia*) and ash F (*Fraxinus*) also correspond with predictions but pine P (*Pinus*) and herbs H (dotted line) seem to depart from this pattern at least at certain times. (V and R from Scholander *et al.*, (1965) M.F. from Milburn *et al.* (unpublished) and P from Helkvist *et al.*, 1974).

in transpiring leafy ring-porous trees such as *Fraxinus americana* at mid-day, all supporting the view that in slow moving systems and in ring-porous trees the gradients are almost entirely caused by gravity with a very small effect of friction. Recent evidence from Helkvist *et al.* (1974) indicates that some conifers may have a considerable frictional resistance to flow, since the pressure gradients during transpiration were 2 bar m^{-1} when conditions were sunny and gradients of 1 bar m^{-1} were found regularly under overcast conditions. These results cannot be attributed entirely to gymnosperm tracheidary wood structure, which might be expected to cause frictional losses, because Scholander's results were also obtained from redwoods and firs.

Practically all of the observed gradients which do depart from the expected gravitational influence can be explained simply by frictional resistance to sap flow drawn by cohesion. Observations of 'negative' gradients have been reported (Tobiesson *et al.*, 1971) in *Sequoiadendron*. One likely explanation, noted by Scholander *et al.* (1965) is that over-trimming shoots can release tensions and produce incorrect readings.

Another way to express the trend of current investigations is in terms of the maximum tensions in the upper parts of plants. Several results shown in Table 5.4 illustrate the fact that they are remarkably similar – irrespective of the plant height under normal environmental fluctuations.

The extensive amounts of conducting xylem in leaves and stems has led at times to the suggestion that many plants are overvascularised.

Table 5.4 Xylem sap tensions generated in upper parts of plants in relation to height under normal environmental conditions under field conditions.

Plant	Height of plant (m)	Max. sap tensions field observation (bar)	Workers
Redwood tree, *Sequoia*	82	*c.* 15.5	Scholander *et al.* (1965)
Coconut palm, *Cocos*	4.5	*c.* 10	Milburn and Zimmermann (1977)
Castor bean, *Ricinus*	1	*c.* 8	Milburn and Dodoo (unpublished)
Wheat, *Avena*	*c.* 0.1	*c.* 12	Passioura (1972)
Plantain (weed) *Plantago*	*c.* 0.01	*c.* 11	Milburn and McLaughlin (1974)

Another explanation is that certain types of plant, especially tall trees, must protect themselves from cavitation effects because they are extremely difficult to reverse except by new growth (but see Ch. 9). In contrast are small herbs which seem to suffer from remarkably great xylem tensions as indicated by cavitation studies (Milburn and McLaughlin, 1974), calculations of flow rates in wheat (Passioura, 1972) and also water potential studies of leaves of many herbaceous crops growing with plentiful water supplies. Apparently many of the smaller herbs (e.g. *Plantago*, *Gramineae*, *Brassica* spp) are capable of restoring xylem sap continuity by the operation of root pressure and guttation which can be observed regularly in many species (see also Ch. 9).

Detection of cavitation

Cavitation is important because it marks the disruption of sap conduction through a conduit, a process which often seems to be irreversible. The sap is disrupted by the development of a gas bubble at near vacuum pressure. It is possible to observe cavitation in small structures like spores or sporangia microscopically, but observation of the xylem requires surgery which can introduce air through mechanical disruption. Owing to the large number of conduits involved it is difficult to distinguish between these effects.

Milburn and Johnson (1966) introduced an acoustic technique to allow detection of cavitation in whole organs. In its simplest form an excised organ (usually a leaf) is impaled by a microphonic probe. Vibrations are produced when conduits cavitate which may be detected, much as earthquakes are detected, seismologically. When the vibrations are amplified through headphones or a speaker the sounds are 'clicks' with a dominant frequency around 500 Hz. Cavitation in vessels can be monitored acoustically and microscopically. When a click is heard, the walls of the xylem, strained inwards by sap at negative pressure, recoil as a bubble forms instantaneously (Fig. 5.9). The volume of this bubble corresponds with the increased volume of the relaxed conduit. Next the bubble enlarges more slowly. Sap is withdrawn from the conduit, in which Ψ_p is now -1 bar relative to atmospheric pressure, by adjacent conduits still at negative pressures much lower than -1 bar. As cavitation proceeds the aqueous continuum in the xylem is progressively disrupted. A single break in a

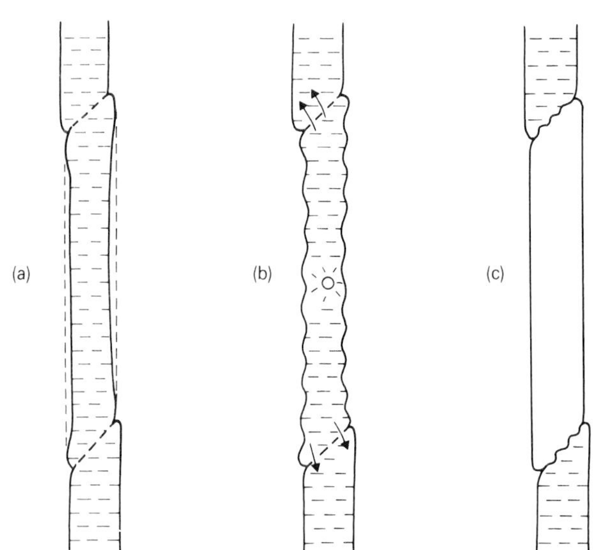

Fig. 5.9 Graphic representation of cavitation in a xylem conduit (*a*). Under tension sap strains the walls inwards from dotted line. (*b*) cavitation 'bubble' forms and strained walls produce vibration detectable as a 'click'. Water exits to adjoining conduits (*c*). Evacuated conduit gradually becomes air-filled by dissolution of gas from sap in walls, etc.

chain of conducting units would prevent flow, but owing to lateral connections rerouting occurs following cavitation limiting the damaging effects when on a restricted scale. Air gradually diffuses into cavitated conduits over several hours, so that while newly cavitated conduits will be restored rapidly if supplied with water ($\Psi=0$) this process becomes much slower with time.

Many experiments have been performed to investigate the effects of environmental factors on cavitation. While a water supply stops cavitation, measures such as radioactive bombardment do not produce detectable increase in cavitation (Milburn, 1973b). It seems clear that cavitation is more readily detectable from large rather than small conduits owing to the greater release of energy from walls during relaxation. Cavitation occurs at surprisingly low tensions (−5 to −20 bar in *Ricinus*, see Ch. 9), much smaller tensions than measured experimentally in physical systems (*c.* −300 bar, see Briggs, 1950) and very much lower than theoretical estimates (−1,000 to −18,000 bar, see App. 5). It seems possible that cavitation may be induced by enlargement of very small bubbles in conduits or by dragging air through larger pores in the conduit walls, a process called nucleation, but as yet this is unknown.

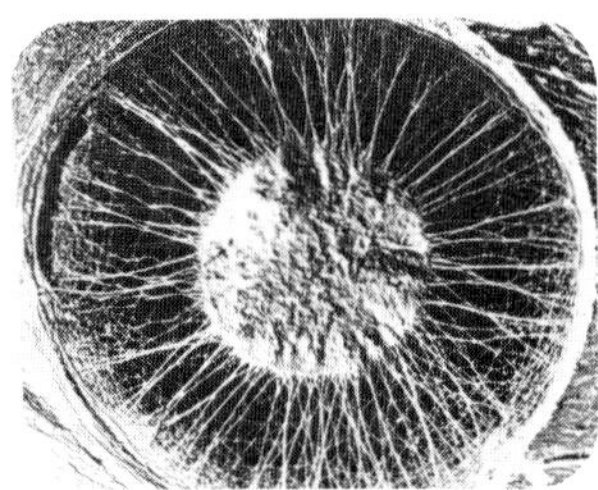

Chapter 6

Water transport in leaves

A typical leaf is a highly refined product of evolution. Not only is it a combination of tissues, i.e. groups of cells with particular teamwork to perform, but their arrangement, which governs the form of leaves, may vary considerably. Conduction systems and devices for controlling water loss, e.g. stomata, hairs or waxy coatings, vary considerably from species to species and even from leaf to leaf on the same plant. We do not yet fully understand how all the systems operate, though any feature retained in the face of evolutionary pressure for many millennia is likely to be connected with some direct or indirect advantage to a plant.

The simplest multicellular plant was probably an alga with cells connected end-to-end in files. No support was required because the system floated in the primeval seas, and even when the form evolved into bilamellar fronds the situation was not radically altered. Seawater provided not only support but also thermal insulation, dissolved carbon dioxide or bicarbonate ions and an amenable water potential; the osmotic potential Ψ_s of present day seawater ranges from -20 to -30 bar but was probably less in ancient seas, because rivers have accumulated salts progressively. However, the conquest of land demanded drastic changes of form, eventually leading to the development of roots, stems and leaves, which enable it to survive desiccation, support itself and yet intercept sunlight.

Typical leaf

A *Ricinus* leaf is typical of many *mesophytes*, which make up the bulk of our crops. Many *xerophytes* living in arid environments, *hydrophytes* living in aquatic environments or *halophytes* living in saline environments, are specialised to combat their particular environment. *Ricinus* is relatively unspecialised yet remarkably capable of survival in both xeric and saline conditions, being physiologically very tough.

The main veins and petioles are invested with tough supporting tissues (collenchyma). Inside is a matrix of parenchyma cells which provide power for growth by turgor expansion. The relationship between compression within and tension outside is a finely regulated system of mechanical support, because the leaf requires support from turgor pressure of the vein parenchyma during growth. When the turgor pressure falls the leaf becomes flaccid, a condition called *wilting*. Simple experiments with scissors quickly reveal that the ribs support the lamella which is suspended between the midribs. The lamella consists of an epidermal sandwich filled with photosynthetic cells. The epidermal cells are like pieces of jigsaw, flattened, with convuluted interlocking edges. Walls are often thick and cuticularised to the exterior with stomatal pores at intervals each bounded by a pair of guard cells able to regulate the porosity (see Fig. 6.1).

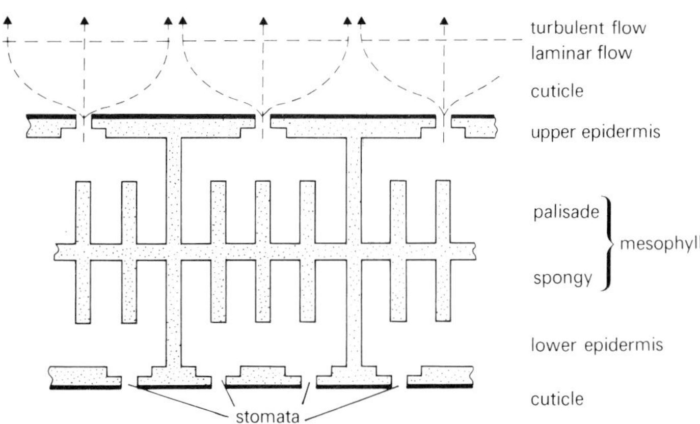

Fig. 6.1 Diagram of a leaf in transverse section showing upper and lower epidermes protected by cuticle. Palisade and spongy mesophyll are represented as gas-exchange fins protected by epidermal layers bearing stomata from desiccation. The arrows indicate how the diffusion path widens above the stomatal pore increasing the apparent efficiency of transport to a distance at which vapour shells mutually interfere.

The photosynthetic cells are of two types, an upper layer of sausage-like palisade cells slung from the upper epidermis and a lower spongy layer of mesophyll. The whole interior is filled with air spaces rendering the gas exchange surface up to forty times greater than that of the leaf itself.

In the veins are two major conducting systems. Xylem carries solutes in the transpiration stream, but its main function is the irrigation of the leaf tissues. The xylem conduits must resist sap tensions but also permit elongation. Strengthening is derived from lignin spirals (see Fig. 5.1) which can extend as growth proceeds. The danger of leaf damage is considerable, e.g. from wind, predators, such as caterpillars, etc. Excessive embolisation of the xylem is prevented by tracheidary conduction, so water must traverse the gas-stopping pit membranes at frequent intervals. The phloem transport system, consisting of fine-bore sieve tubes with very large companion cells, carries organic solutes as a bulk flow of sugary sap, predominantly in the opposite direction. Both conduction systems are heavily invested with fibrous cells which support and protect them mechanically.

Beyond the veins proper sap conduction proceeds on a cell to cell basis. Veins are normally invested in layers of living parenchymatous

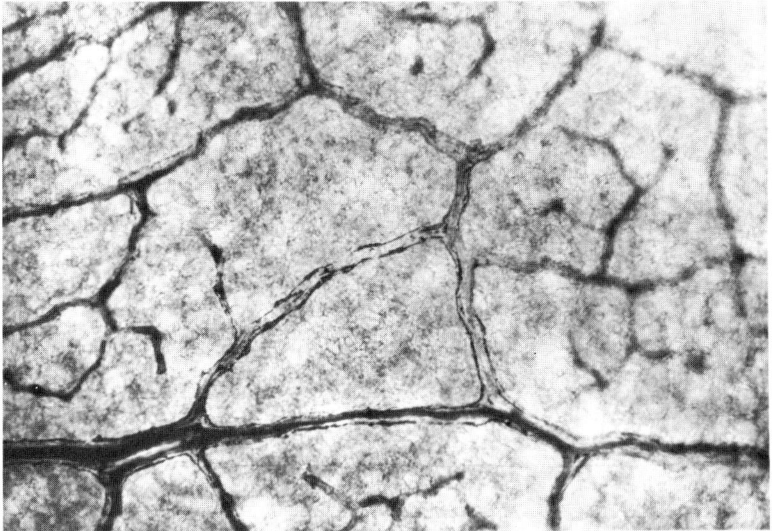

Fig. 6.2 Illustrating vein extension cells heavily stained with lead sulphide fed into the xylem as lead chelate and fixed with H_2S. The cleared leaf is seen from above and the intricate branching of the minor veins is obvious surrounding mesophyll 'islets' of cells. (*Scale:* 1 cm = 3 μm)

cells, sometimes extending in processes towards the epidermes called *vein extensions*. From experiments with dyes and tracers it seems that these cells are unusually permeable to water and are probably specialised for this purpose in having (see Fig. 6.2) *exceptionally* porous cell walls. Significantly, vein extensions connect with the epidermal layers of cells and there is growing evidence that epidermal cells provide an important water distribution system to the enclosed photosynthetic cells.

Between the complex network of veins, the mesophyll cells can be considered as islets of cells. In paradermal section the number of cells across each islet seldom exceeds ten, which implies that the xylem water supply to the innermost cell would have to cross only three to five cells, the number of cells in the mesophyll catena (chain).

Leaf and mesophyll areas and gas volume

The most logical reference area to be considered when studying leaf physiology is the area which intercepts radiant energy, which we will call the leaf area A_r, because this limits the capacity of a leaf to photosynthesise. It also governs the area over which radiant energy is received for the evaporation of water in the process of transpiration. The latter is highly influential, as may be realised from the fact that $25.06 \times 10^{-4}\,J$ are required to evaporate 1 mol of water (18.016 g) at 25°C.

In most leaves we can assume that A_r corresponds with the area of the upper epidermal surface, though a correction factor may be necessary for shading or surface irregularities. In most mesophytic leaves the total area of epidermis from which water can evaporate $A_e = 2A_r$. Since water loss is governed more by stomatal distribution and energy influx than actual area, the measurement of A_e may not be particularly relevant. Some herbs have stomata in both upper and lower epidermis; in trees the stomata are often present only on the lower epidermis, and in hydrophytes on the surface exposed to air (the upper epidermis). Areas are often measured by drawing the outline of a leaf and determining the weight of the cut out in relation to the weight and area of the sheet of paper. Alternatively the outline may be traced with a planimeter. Photographic or light-sensitive paper methods have been used. For more rapid estimation of leaf areas it is possible to measure areas photometrically and even electronic scanning techniques have been used. An indirect approach is to measure leaves using one or more linear measurements and relate this area using a calibration graph so that subsequently area measurements can be made with a simple ruler. Alternatively a correlation between weight and area of leaves is established. However, leaf weight per unit area is strongly subject to the level of illumination

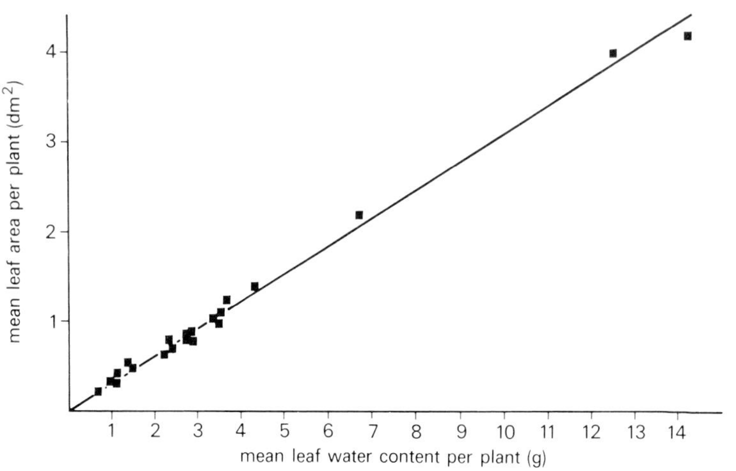

Fig. 6.3 Mean leaf area per plant – mean leaf water content of field-grown *Helianthus annuus* showing a linear relationship (from Evans, 1972).

during growth so this method must be used with caution (see Fig. 6.3).

The internal surface area of a leaf A_i is vastly greater than its superficial area A_r. Surface areas can be measured from thin transverse sections of a leaf by measuring the linear outline of cells at a particular focal plane in relation to the epidermal surfaces visible. (This method depends on having a short focal distance and is most accurate using high power objectives.)[1] Some xerophytes have a smaller internal surface than mesophytes. In mesophytes A_i may be as high as 20 A_r. In many respects the internal structure of a leaf resembles that of an animal lung which is also designed for rapid gas exchange – the average surface area of lungs in a man is a staggering 55–70 m^2!

The volume of gas space within a leaf is very considerable, ranging up to about 50 per cent of the whole leaf. Such measurements can be performed by measuring the weight increase of a leaf after it has been injected with water under vacuum.

Cellular transport in leaves

The previous anatomical description makes it clear that the pathway taken by water in a leaf is liable to be complex. Water seeks a path of least resistance between two points at different water potentials; so how much do we know quantitatively about the conductivity of the alternative pathways? Liquid water is unlikely to be transported via

[1] Correction factors may be needed for oblique walls.

protoplasts because of the high resistance to flow across the plasmalemma (and tonoplast?) which must be penetrated twice each time a cell is traversed (see also Ch. 8). In contrast the cell walls are undoubtedly permeable to water since they have been shown to transport dyes, lead chelate and colloidal gold from the xylem to the epidermal cells. These tracers have relative molecular masses exceeding 400 which must imply free access to much smaller water molecules like water.

There must be a frictional component within such fine capillaries (see Fig. 6.4) which is bound to alter the Ψ_p gradient sharply over short distances in comparison with xylem transport. On drying cell walls shrink which implies that capillaries within also shrink in diameter – larger capillaries may even become air-filled (producing hysteresis effects as described in soils). Vein sheath, vein-extension and epidermal cells have probably a greater hydraulic conductivity than other mesophyll cells, but mesophyll cells have in general thicker walls than ordinary parenchyma cells.

The length of the mesophyll catena is at most about five cells (in *Ricinus* the number is only about three). The number of cells adjacent to vascular tissues is much greater than the single 'terminal' cell in the centre of each islet. Assuming the cells are circular in cross-section and equal in size, the number of cells across the path from the innermost cell will be 1, 6, 12, 18 and 24 in number. The volume of cell walls in *Ricinus* mesophyll is about 12 per cent of the cell volume (from EM sections) so if the volume of water in the walls is about half of this there will be a maximum of 6 per cent of the cell water in cell walls. The total distance is only about 17 μm to the central islet cell but 50–60 μm from upper to lower epidermis.

There is little firm evidence on the magnitude of the three possible paths of water transport across the mesophyll catena. It is most likely that liquid transport is mainly via cell walls and only a small fraction

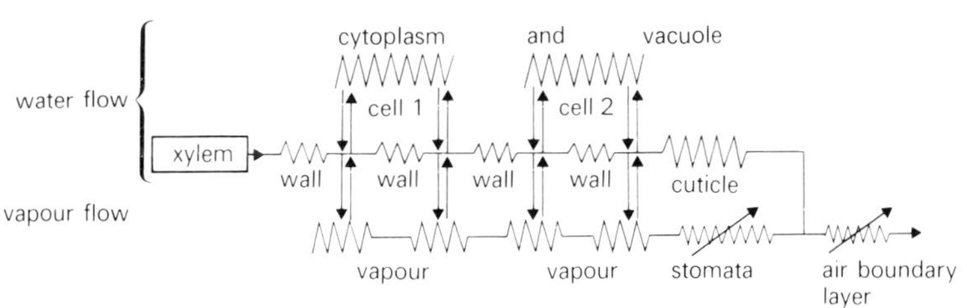

Fig. 6.4 Representation of a simplified mesophyll catena of only two cells to illustrate the various pathways series and parallel through which water is conducted. Protoplasts are peripheral to the main pathway. All the resistances are variable but the main variable resistances are depicted as rheostats.

is transported via protoplasts (see also Ch. 8). Distillation of water across the gas phase between mesophyll and epidermal cells could be important however, depending upon the magnitude of internal thermal gradients set up within the leaf.

Transpiration J_{wv}

The earliest land plants could only survive in very moist situations, and a further conquest of land required a piped supply of irrigation water via xylem itself supplied by an efficient collection system – the root. This problem was still further aggravated by the fact that photosynthetic cells had lost their supply of dissolved carbon dioxide. Instead, a plant needed to expose the delicate photosynthetic cells to the ravages of a desiccating atmosphere. Atmospheric desiccation was a persistent problem because even when the atmosphere was half saturated with water (50 per cent r.h.) its water potential approached –1,000 bar (936 bar at 20°C). An extremely porous system was developed, enveloped in protective lamellae, the epidermes. Gas permeability could be varied by the action of stomata. In this way a compromise was achieved between the need for maximum exposure of internal surface for gas exchange (carbon dioxide in air is now only about 0.03 per cent by volume but this concentration may be lower than formerly) and the maximum interception of sunlight.

The evaporation of liquid water into the vapour phase from exposed plant organs is called *transpiration*. The current of sap drawn through the plant in response to transpiration is called the *transpiration stream*. Transpiration is not easy to measure without actually modifying the system. Potted plants can be weighed regularly and the loss of water computed, but this method cannot be applied in field studies. Attempts to measure transpiration by enclosing a leaf usually alters the environment significantly. The simple method based on detaching a leaf and measuring the rate of water loss by change in weight before the stomata have time to change is open to similar objections. If a shoot is fitted to a potometer (Fig. 6.5) transpiration is indicated but only under steady conditions when the cell water balance is constant. Water potential of the potometer supply is also abnormally close to zero.

One modern technique used for fieldwork uses meteorological measurements. Wind speed and changes in humidity are computed to give the water loss from a stand of vegetation. Alternatively, transpiration from plants and leaves can be estimated using a diffusion porometer (Fig. 7.3). Another technique is the *lysimeter*, a container in which even trees can be grown in large masses of soil in much the same way as potted plants. Water balance from irrigation,

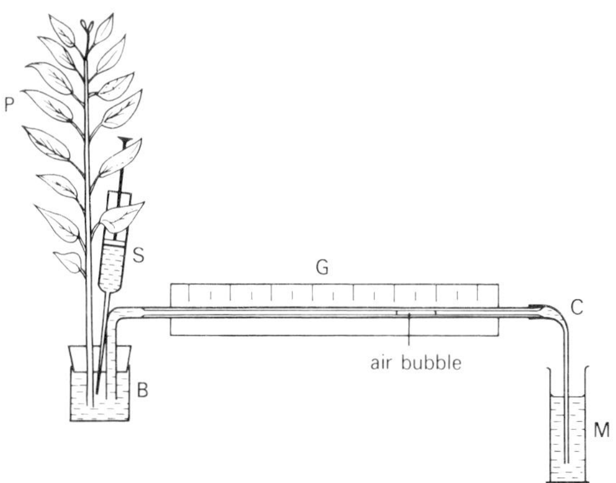

Fig. 6.5 A simple potometer in which a plant shoot P is sealed in a rubber stopper in a small glass bottle B. Uptake is measured by the movement of the airbubble across graduated scale G and the bubble is rezeroed using syringe S. Prolonged uptake is measured when the bottle is expelled and attachment C fitted to take up water from measuring cylinder M.

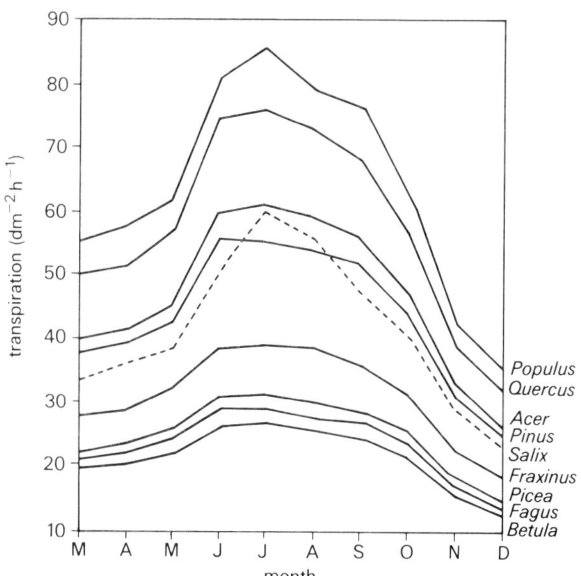

Fig. 6.6 Transpiration rates J_{wv} of common trees in temperate regions throughout the season. All show a mean transpiration rate related to the net radiation which is greatest in the summer months (from Gindel, 1973).

transpiration and rainfall is computed by weighing the container automatically at regular intervals. Transpiration, J_{wv} is expressed in different units depending on the application (see Ch. 1). Typical values are presented per unit surface area of leaf A_e in Table 6.1 and Fig. 6.6.

Transpiration does not seem beneficial from a physiological point

Table 6.1 Transpiration rates calculated per area of leaf surface A_e from mesophytes and xerophytes over brief periods and daily rates. Rates represent the maximum rates expected under sunny warm conditions. Xerophytes have g[illegible]erally lower transpiration rates than mesophytes but as shown by salt-steppe me[illegible]urements this is not automatically the case (Note: $J_v \approx J_{wv} \times 10^{-3}$ m s^{-1}).

Conditions of measurement	Plant genus	Rate of transpiration J_{wv}		Wo[illegible]er and ye[illegible]
		g m^{-2} h^{-1}	$\times 10^{-5}$ kg m^{-2} s^{-1} (= μg cm^{-2} s^{-1})	
Over short periods				
Maximum rates	*Eucalyptus*	750	20.8	Wilson (1924)
	Helianthus	440	12.2	Martin and Clements (1933)
Mesophytes (typical)	–	250	6.9	–
Xerophytes (typical)	–	25	0.7	–
Salt-steppe plants	*Statice*	246	6.8	Stocker (1933)
Salt-steppe plants	*Artemesia*	264	7.3	Stocker (1933)
Over 24 hour periods				
N. Carolina, U.S.A.	*Ilex*	67	1.86	Kramer (1969)
N. Carolina, U.S.A.	*Quercus*	59	1.64	Kramer (1969)
N. Carolina, U.S.A.	*Liriodendron*	49	1.36	Kramer (1969)
N. Carolina, U.S.A.	*Pinus*	21	0.59	Kramer (1969)

of view. The transpiration stream appears to assist in salt uptake only when salts are already abundant. Though transpiration undoubtedly provides cooling, leaves seem better able to withstand high temperatures than desiccation. It is reasonably regarded as an 'unavoidable curse' though it may play a minor role in facilitating the circulation of solutes within plants.

Van den Honert's analysis

An important advance in the understanding of water movement in leaves is due to Van den Honert (1948) based on an idea of

Gradmann. It is argued that in a plant transpiring at a constant rate over an extended period all parts of the plant would have reached equilibrium and the flow through root J_r, stem J_s, leaf J_e, and gas J_g, would be equal. Under these conditions

$$J_r = J_s = J_l = J_g \qquad [6.1]$$

Since the flow rate must be driven by a water potential gradient the magnitude of this gradient must reflect the hydraulic conductance L_p of the interlinked components so that

$$(\Delta\Psi L_p)_r = (\Delta\Psi L_p)_s = (\Delta\Psi L_p)_l = (\Delta\Psi L_p)_g \qquad [6.2]$$

If approximate values are attributed to the water potential in root, stem, leaf and gaseous systems the hydraulic conductance L_p of the components is revealed. Thus if the water potential of soil is around −1 bar, and of leaf cells is about −20 bar, only 19 bar have been utilised to draw water from soil to the leaves through a path which includes the stem. The air outside the plant has a relative humidity of

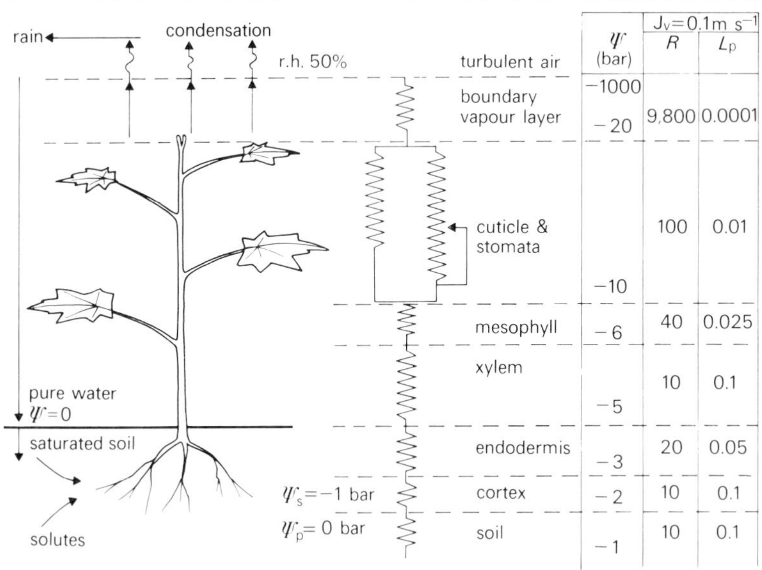

Fig. 6.7 Illustration of the Van den Honert analysis showing conjectural water potentials at different parts of a plant between well watered soil and the gas phase. Centrally is an electronic analogue of the plant with (*right*) numerical figures for Ψ and J_v (0.1 m s^{-1} throughout) on the basis of which L_p, and its reciprocal R, are computed. Both are in terms of water potential (Ψ bar). The conductance at the gas phase is the greatest limitation to water transport in the catena. *Note:* Hydraulic resistance $R \neq$ gaseous resistance R (see page 116).

only 50 per cent, equivalent to about −936 bar at 20°C, a severe reduction in hydraulic conductance somewhere between the cells of the leaf and the atmosphere is implied (see Fig. 6.7). This low conductance barrier could not be situated *within* the leaf, otherwise all tissues beyond it must be subjected to severe desiccation. The only tenable conclusion is that the barrier to conduction must lie beyond the living plant tissues themselves at or in the gas phase.

Van den Honert argued that this barrier must lie in the stomata and especially in the boundary layer of unstirred air which normally clings to plants. In the gaseous phase, the flow of water vapour is slowed by the low diffusion coefficient of water in air which in turn reflects the hindrance to the free diffusion of water molecules imposed by gas molecules.

The regulation of transpiration by stomata

Much of the surface of a plant exposed to air is cuticularised so preventing excessive loss of water. This insulation would drastically reduce the gas exchange essential for photosynthesis were it not for the presence of stomata. Each stoma consists of a pair of guard cells surrounding the stomatal pore through which gases diffuse. Apparently during their evolution plants have been unable to permit gas exchange without permitting the escape of water vapour, hence the necessity of the xylem 'irrigation system'.

Though the guard cells may occupy as much as 20 per cent of the epidermal surface of a leaf the stomatal pore area is much smaller. The area of all the stomatal pores is usually in the range 0.5–5 per cent and commonly about 1 per cent of the leaf area A_e (see Ch. 7). The pores are minute but extremely numerous (see Table 7.1), consequently their capacity to permit diffusive gas exchange when open is remarkably high (Fig. 6.1). Why do leaves require stomata instead of, say, an array of minute fixed perforations? The answer is not obvious because much of the time stomatal control is not essential and simple perforations would indeed suffice.

Despite the fact that a leaf may be surrounded by air, often with a water potential Ψ around −1,000 bar, Ψ of the mesophyll cells inside the leaf is usually −10 to −30 bar. A barrier with fixed pores could limit the escape of water vapour because a diffusion gradient would be established between the pore and the atmosphere limiting the rate of diffusion very considerably. The problem lies in the fact that such a system, with a long diffusion pathway in the boundary layer established in still air, would be changed drastically when exposed to wind. In wind, boundary layers of air partially saturated with water vapour which clings to the plant surfaces, are largely blown away by turbulent air flow leaving much thinner boundary layers. Since the diffusion of water vapour increases if the path length is shortened, the transpiration rate would increase dramatically. Consequently stomatal closure allows the plant to survive the deleterious effects of wind or intense radiation (though this control may not always be adequate if changes occur very suddenly). Nevertheless stomata can open in still air, so that gas exchange for photosynthetic assimilation is maximised far in excess of that possible through the minute fixed pores which would be necessary to withstand desiccation in wind.

The paths of gaseous diffusion

It is commonly assumed that the escape of water vapour is matched by a corresponding influx of carbon dioxide and an efflux of oxygen, an excretory product from photosynthesis. In fact the water vapour pathway is more extensive and also shorter than that for oxygen and carbon dioxide. Much water escapes from epidermal cells which lack photosynthetic pigments but nevertheless absorb infrared radiation. Stomata seem unable to export assimilates which alone must limit their photosynthetic capacity. Water vaporises from the *surfaces* of mesophyll cells but carbon dioxide must penetrate the walls in solution and pass through both plasmalemma and cytoplasm to reach the chloroplasts. Oxygen escapes via the same path during photosynthesis. Air is unable to penetrate the cell wall capillaries, despite the sap tension (Fig. 6.8), on account of the surface tension of water.

Carbon dioxide gas is at a very low concentration in normal air, around 0.03 per cent by volume, or even less, so the diffusion gradient through the stomatal apparatus cannot be very great and its high relative molecular mass gives it a lower diffusion coefficient than either water vapour or oxygen (see App. 3). When I see a forest fire I cannot but marvel to appreciate that every scrap of burning carbon has been gathered as dilute carbon dioxide gas. Each gram of carbon corresponds with the extraction of 6.1 litres of pure carbon dioxide gas (at STP) from over 20,000 litres of normal atmospheric air. Such

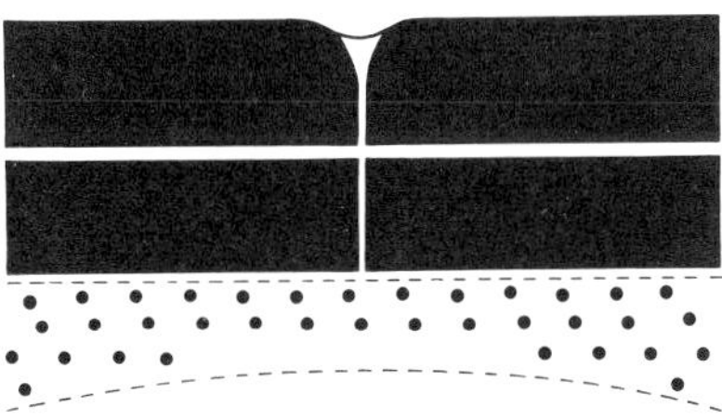

Fig. 6.8 Model to illustrate water flow through a cell wall involving tangential and radial channels with respect to the cell protoplast. Evaporation occurs from the exposed meniscus which prevents the entry of air by surface tension.

is the efficiency of the gaseous exchange systems in plants; it is little wonder that such a mechanism necessitates the transpiration of considerable quantities of water vapour!

Resistance of stomatal and boundary gas layers

The loss of water vapour from leaves is limited by diffusion either in stomatal pores themselves or in boundary layers which cover leaf surfaces generally. The former is controlled internally by the guard cells of the plant; the latter by changes in the external environment. Hence the concept arose of measuring the diffusive resistance of the leaf surface composed of these two components. Later the need arose to separate these measurements.

An early technique for measuring diffusive resistance was introduced by Francis Darwin (1898) who placed strips of hygroscopic horn against leaf surfaces. The degree of horn curvature, measured over a fixed time, indicated the rate of water loss from the leaf. This is the basis of modern diffusion porometers in which horn strips are replaced by electronic humidity sensors. The instruments are calibrated in terms of evaporation at a free water surface.

The conventional procedure used to study the diffusive resistance to water transport from leaves has been derived by Bange and others. Fick's law for diffusion is approximated to the linear form thus:

$$J_{wv} = D\frac{dc}{dx} \cong \frac{D\Delta c}{x} \qquad [6.3]$$

Where J_{wv} is the diffusive flux of water vapour (kg m^{-2}), D is the diffusion coefficient of water vapour in air (see App. 3) and c/x is the vapour concentration gradient over path x. The expression is modified to a form analogous with Ohm's law so that:

$$J_{wv} = \frac{\Delta c}{R} \qquad \text{Therefore } R = \frac{\Delta c}{J_{wv}} \qquad [6.4]$$

Where R is the gaseous diffusive resistance in units of s m^{-1}. In practice R is measured by comparing the diffusive resistance of a leaf with an equivalent exposed surface of water, for example a moist absorbent paper of the same shape and properties as the leaf and under the same conditions (see Holmgren, Jarvis and Jarvis, 1965). Values for R have been calculated for boundary air and cuticular and stomatal paths (Table 6.2).

Table 6.2 Resistances R to water vapour transfer of light-saturated leaves of trees and herbs at 22°C (from Holmgren, Jarvis and Jarvis, 1965).

Plant type and species		**Resistances to water vapour transfer** ($s\,m^{-1}$)		
Type	**Species**	**Cuticle** R_c	**Stomata (open)** R_s	**Gas phase** R_a
Herbs	*Helianthus annuus*	–	38	55
Herbs	*Lamium galeobdolon*	3,700	1,060	73
Herbs	*Circaea lutetiana*	9,000	1,610	61
Trees	*Acer platanoides*	8,500	470	69
Trees	*Quercus robur*	38,000	670	69
Trees	*Betula verrucosa*	8,300	92	80

The reasons for using these units are as follows:

1. In the gas phase water transfer is directly proportional to the concentration gradient of water vapour which is *not* linearly related to gradients of water potential.
2. Resistances are more easily manipulated than conductances when considering transpiration as a catenary process.
3. Diffusive resistances are used to measure gas fluxes associated with photosynthesis and respiration.

In combination diffusive resistances can be treated like electrical resistances. In series they are added together – in parallel the flow is shared on a fractional basis (see App. 15). If the flow from mesophyll cells into turbulent air is considered water must traverse the stomatal resistance R_s and cuticular resistance R_c which are in parallel. Both of these resistances are in series with the resistance of the boundary layer of air R_a. The diffusive resistance of the leaf surface R_l is as follows:

$$R_l = R_a + \frac{R_s R_c}{R_s + R_c} \qquad [6.5]$$

The diffusive resistance of still air is a fixed quantity at a given temperature. An important consequence of Equations [6.3] and [6.4] is that R_a of a 1 m column of air is the reciprocal of the diffusion coefficient D of water vapour in air. At 22°C D equals 3.865×10^{-5} $m^2\,s^{-1}$ giving R_a a diffusive resistance of 25,873 $s\,m^{-1}$. Consequently if R_a equals 60 $s\,m^{-1}$ (see Table 6.2) this implies a boundary layer of unstirred air of 60/38,650 m = 1.55 mm. Similarly if the mean depth of a stomatal pore is 10 μm and R_s also equals 60 this implies a reduced cross-sectional pathway (the combined cross-sectional area of stomatal pores) of 10/1,550 μm = 0.6 per cent. The legitimacy of the latter calculation may be questioned however, because diffusion coefficients tend to lose their legitimacy at this microscopic scale on account of 'edge effects'. Increasing use is made of diffusive resistance in reciprocal form, $1/R$, the diffusive conductance, which resembles hydraulic conductance (see below).

The hydraulic conductivity of the transpiration pathway

Hydraulic conductivity is normally applied to the movement of *liquid* water but we can treat flow through the gas phase *as if* it were through an elastic porous matrix of still air resembling flow through a soil or a cell-wall matrix. In this way we can consider the water transpired as driving an equivalent hydraulic flow through the whole plant under steady conditions, i.e. the transpiration stream. Transpiration J_{wv}, measured as weight of water lost per square metre of leaf surface, can be converted into our familiar J_v, which is the volume of water per unit area per unit time. Therefore $J_v = J_{wv}/\rho$ the density of liquid water (see Ch. 1).

$$J_v = L_p \Delta\Psi = L\frac{\Delta\Psi}{x}$$

Where L_p is the hydraulic conductance and L the hydraulic conductivity of the gas phase (air) and Ψ is the water potential gradient over path x. The problem now is to incorporate effects of windspeed, thermal gradients and the hydraulic conductivity of the gas phase into the equation so that we may measure the variables accurately.

Hydraulic conductivity of the gas phase

In still air Fick's equation applies to the diffusion of water vapour with reasonable accuracy. Thus

$$J_{wv} = D\frac{\Delta c}{x} \quad \text{But } J_{wv} = J_v \rho \quad \text{and } J_v = L\frac{\Delta\Psi}{x}$$

From these we can deduce that

$$L = \frac{D}{\rho} \cdot \frac{\Delta c}{\Delta\Psi} \qquad [6.6]$$

where D is the diffusion coefficient of water vapour in air and ρ the density of liquid water. Δc and $\Delta\Psi$ refer to the water vapour concentration and water potential gradient across the same gas phase. Δc can be expressed in terms of the relative humidity (r.h.) of air from which Ψ can be deduced. Thus $\Psi - \Psi' = \Delta\Psi$ where Ψ and Ψ' are different water potentials across the gas phase which can be calculated:

$$\Delta\Psi = \frac{RT}{V}\ln\left(\frac{c_o}{c_{sat}}\right) \times 10^{-5} = \frac{RT}{V}\ln\left(\frac{\text{r.h.}\%}{100}\right) \times 10^{-5}\ \text{bar} \qquad [6.7]$$

Water vapour content of air saturated	c_{sat}	ML^{-3}	$kg\,m^{-3}$
Water vapour content of air part sat.	c_o	ML^{-3}	$kg\,m^{-3}$
Water potential of the air	Ψ	$ML^{-1}T^{-2}$	$N\,m^{-2}$ (= Pa)(= bar × 10^{-5})
Relative humidity	r.h.%	$ML^{-3} \times M^{-1}L^{-3}$	$\frac{\text{conc. water in air}}{\text{conc. in sat. air}} \times \frac{100}{1}$
Gas constant	R	$ML^2T^{-2}\,mol^{-1}\,K^{-1}$	$J\,mol^{-1}\,K^{-1}$
Temperature	T	K	K
Partial molal volume of water	V	$L^3\,mol^{-1}$	$m^3\,mol^{-1}$

From these equations L the equivalent hydraulic conductivity of a 1 m column of air can be related to both relative humidity of air and the water potential as shown on Fig. 6.9. It can be seen that the effect

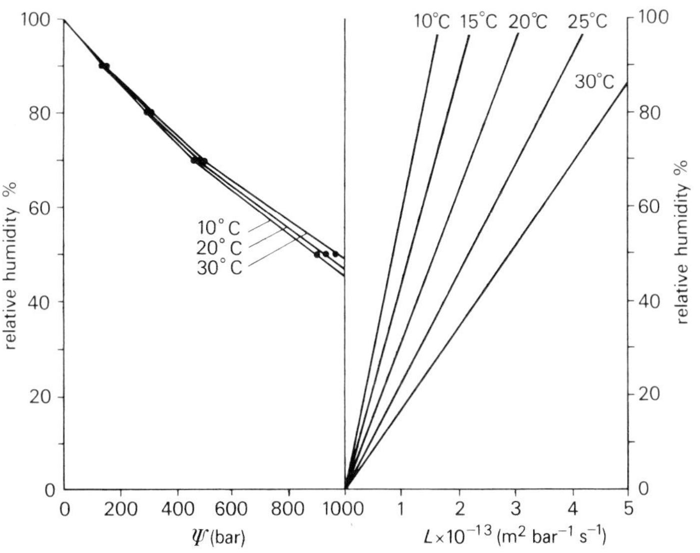

Fig. 6.9 Graphs relating water potential with the relative humidity of the air and its apparent hydraulic conductivity L. The effect of temperature on L is considerable and over a thermal gradient a mean value for L must be computed.

of temperature is rather small on Ψ but it has a considerable effect on the value of L.

Thickness of the boundary gas layer

The thickness of the boundary gas layer is difficult to measure in experimental conditions but it can be deduced from the equation below derived from wind tunnel experiments. Nobel (1974) gives a value for the Factor of 0.4: other values have been suggested:

$$x = \text{Factor} \times 10^{-2}\left(\frac{l}{V}\right)^{\frac{1}{2}} \qquad [6.8]$$

Here x is the equivalent boundary layer thickness, l the leaf dimension parallel to the wind direction (m) and v is the wind velocity over the leaf ($\mathrm{m\,s^{-1}}$).

From Equation [6.8] we can compute typical values for the boundary layer over a $(0.1\ \mathrm{m})^2$ leaf under different conditions as shown in Table 6.3. (It may be noted that in practice 'still' air has a

Table 6.3 Examples showing parameters calculated for water vapour transfer via the diffusive gas phase from a $(0.1\ \mathrm{m})^2$ leaf subjected to different wind speeds at 15°C (40 per cent r.h.). The leaf surface is at 15°C and near 60 per cent r.h. Diffusive path length x is calculated from Nobel's formula. If the water potential difference $\Delta\Psi$ across the gas phase and the hydraulic conductivity L of the air remain constant, then a reduction in wind velocity V must increase x. As the diffusive path length x increases the transpiration rate J_v is reduced as shown; alternatively the effective $\Delta\Psi$ could be reduced by raising the water potential at the leaf surface. The apparent increase in hydraulic conductance L_p as wind velocity is reduced towards zero is entirely caused by changes in x.

Water potential difference $\Delta\Psi$ bar	**Wind velocity** V $\mathrm{m\,s^{-1}}$	**Diffusive path** x m	**Transpiration rate** J_v $\mathrm{m^3\,m^{-2}\,s^{-1}}$	**Hydraulic conductance** L_p $\mathrm{m\,s^{-1}\,bar^{-1}}$	**Hydraulic conductivity** L $\mathrm{m^2\,s^{-1}\,bar^{-1}}$
538	10.0 (high)	0.40×10^{-3}	16.1×10^{-8}	3.0×10^{-10}	1.2×10^{-13}
538	1.0 (typical)	1.26×10^{3}	4.8×10^{-8}	0.9×10^{-10}	1.2×10^{-13}
538	0.1 (still)	4.00×10^{3}	1.6×10^{-8}	0.3×10^{-10}	1.2×10^{-13}

velocity approaching $0.1\ \mathrm{m\,s^{-1}}$.) The most accurate device for measurement of steady windspeeds is the hotwire anemometer. Wind speeds of $10\ \mathrm{m\,s^{-1}}$ reduce the thickness of the boundary layer from 4 mm thick to a mere 0.4 mm. Recently Sherriff (1973) has used an infrared detection device to measure boundary layers directly.

Water potential gradient (isothermal)

It is relatively straightforward to determine the water potential of air beyond the boundary layer by placing an hygrometer between the boundary layer and the turbulent air above. In practice the humidity gradient in the turbulent air is slight so there is little error derived from inaccurate positioning.

In principle we could measure the water potential at the leaf surface by placing small hygroscopic crystals or minute droplets of osmotic liquid in close proximity to the leaf surface (Noble-Nesbitt, 1975). At equilibrium the osmotic potential of the droplet balances the relative humidity of the air at the leaf surface and from its volume change or liquefaction the water potential at the leaf surface can be measured. In this way the apparent water potential across the boundary gas layer can be measured directly. However there is an additional component to consider – the thermal gradient.

Water potential gradients (non-isothermal)

Leaves are commonly 1°C warmer or cooler than the ambient air but on occasion leaves have been found to be 7°C or more higher than the ambient air. A temperature increase implies a radiation imbalance which is not compensated by the evaporation of water from leaves. Even a small temperature gradient induces distillation of water across a gradient of water potential, but what is its magnitude?

The water potential gradient induced by a thermal gradient has been determined experimentally to a high degree of accuracy by Stokes (1947) for quite another purpose. Results derived from this paper are shown in Fig. 6.10 (using the bithermal method for studying vapour pressure in which an osmoticum at one temperature is balanced against a cooled liquid, see inset). This indicates that Δ1°C is equivalent to $\Delta\Psi$ of 81.6 bar! (This finding can be used to estimate the tensile strength of water, see App. 5.) An identical value can be derived from Equation [6.7] providing the effect of temperature on r.h. is correctly computed, so we may be sure this value is correct. To measure the temperature of a leaf it is necessary to attach small thermistors or thermocouples to the leaf surface. Air temperatures can be measured using precision mercury thermometers. The water potential produced by a thermal gradient must be added or subtracted from the hygrometric water potential depending on the direction of induced distillation. The final equation therefore can be expressed in the units specified previously:

$$\frac{J_{wv}}{\rho} = J_v = L_p \Delta\Psi = L \frac{\Delta\Psi}{x} \qquad [6.9]$$

where $\Delta\Psi = \Psi$ (isothermal) $+ \Psi$(non-isothermal)

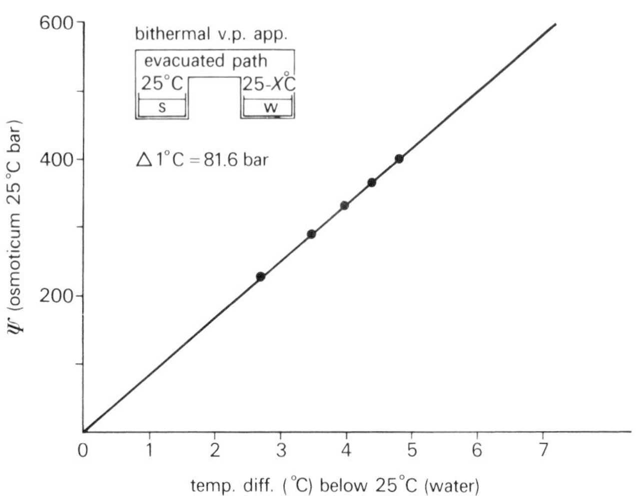

Fig. 6.10 Results from the bithermal vapour pressure apparatus (*inset*) show a linear relationship between Ψ and temperature difference (derived from Stokes, 1947). Distillation is speeded by evacuating the apparatus thus shortening the time to reach equilibrium.

We will now proceed to examine the implications of these findings to estimate anticipated values in simple examples.

Hydraulic conductivity L of the external gas phase

Transpiration of water from a leaf surface is a complex process affected by wind and the input of radiation. It has been examined in Table 6.3 for a typical $(0.1\ \text{m})^2$ leaf, using the simple equations already described. It has been assumed that the water potential gradient across the diffusive layer is constant at -400 bar, which would correspond with a relative humidity in the turbulent air of about 40 per cent ($\Psi \approx -1{,}000$ at 22°C) with the water potential of the leaf surface in the range -600 bar (60 per cent r.h. at 22°C). The hydraulic conductivity, L, of air must be constant (see above); the apparent change in L_p is entirely caused by changes in the mean thickness of the diffusive gas layer. From high wind velocities to still air the transpiration rate J_v from a high value must fall one hundredfold.

One implication of these calculations is that the diffusive gas layer is an unreliable insulator under windy conditions, necessitating effective stomatal control at the leaf surface. Little is known about the water potentials at leaf surfaces. In practice the effect of an increase in wind velocity would be to cool the leaf surface so lowering its temperature, thus reducing the effective water potential gradient. Ideally we should measure this surface temperature directly using very small thermocouples or thermistors. Surface water potential of leaves may be reduced further by local accumulation of sap rich in solutes. Both of these parameters would have to be measured on a transpiring leaf – they would disappear under static conditions, for example in a psychrometer chamber.

Hydraulic conductivity L **of the leaf surface**

Under conditions of rapid transpiration we cannot assume that the water potential above the leaf surface is closely correlated with the leaf water potential Ψ as determined, say, using a pressure bomb because of the large water potential gradient $\Delta\Psi/x$ across the leaf tissues. We do not have exact measurements of the external leaf surface potential; estimates have indicated values down to $-1{,}000$ bar. For simplicity in Table 6.4 it is assumed that during rapid transpiration $\Delta\Psi$ is -600 bar at 22°C to which is added a temperature correction of -400 bar.

Stomata are remarkable because they can transmit water vapour at only about a tenth of the evaporation rate from a free water surface (where $\Psi = 0$ bar), yet the pores occupy only about 1 per cent of a leaf surface when open, $\Delta\Psi$ of the inner leaf surfaces being -10 to -100 bar (see Slatyer, 1967). The customary explanation is that the vapour 'shells' around the pores are separated sufficiently for stomatal pores to be exceptionally efficient conductors, but the most important barrier must be the low hydraulic conductivity of the gas phase in the pore throat. The main reason stomata seem exceptionally efficient probably lies in the radiation balance. A leaf under strong radiation is largely insulated by cuticle and, being unable to lose water over the entire surface, it becomes warmer. A heat flux is set up warming guard cells and the cells supplying water for transpiration more than an equivalent open water surface, so increasing the effective water potential gradient. In Table 6.4 similar values to those used in Table 6.3 are calculated in the usual manner assuming that open stomatal pores occupy 1 per cent of the cuticular surface raising their effective J_v one hundredfold. Even if we assume the stomatal throat is only 20 μm deep, 1,000 bar water under potential difference is necessary to drive flow at the measured rates. A thermal differential of 5°C has been assumed (400 bar/81.6 $\approx$ 5°C) between leaf interior and the external gas phase. The highest temperature differentials measured to date are about 10°C. Perhaps the stomatal throat is much shorter. There is growing evidence that guard cells themselves transpire considerable quantities of water and if this is correct then x can be reduced in Table 6.4.

Finally we may look again at the Van den Honert analysis which

Table 6.4 Calculations showing the water potential differences $\Delta\Psi$ necessary to drive transpiration J_v at the higher rates measured. Leaf temperature is 27°C and the ambient air temperature is 22°C at 60 per cent r.h. Diffusion paths x through the stomata are taken as 20 and 200 μm and through cuticle 0.2 and 2 μm. Stomatal cross-sectional area is assumed to be 1 per cent of the leaf surface. Cuticular transpiration J_v is taken as 1 to 10 per cent of the higher transpiration rate when stomata are open – it has a much greater cross-sectional area than stomatal pores. The very low conductivity of cuticular material is striking. When stomatal transport is high, considerable gradients of water potential, caused by local temperature gradients, are added to vapour concentration gradient.

Transpiration via main component	**Rate rel. transp.**	**Transpiration** J_v $m^3\,m^{-2}\,s^{-1}$	**Diffusion path** x m
Stomata open	High	10^{-5}	2×10^{-5}
Stomata open	Lower	10^{-6}	2×10^{-4}
Cuticular	High	10^{-8}	2×10^{-7}
Cuticular	Low	10^{-9}	2×10^{-6}

gives the impression that the greatest resistance to water flow lies in the gas phase. In still air this is correct, but in wind when the diffusion path has been shortened, the leaf is insulated by the cuticle which per unit thickness is a much more effective insulator than the gas phase (see values for L in Table 6.4). This conclusion is supported by observations that projecting leaf hairs which are often exposed to very severe desiccation survive in a turgid condition despite limited cuticular protection and even do so when severed from the plant. Considering the effective insulation provided by cuticle the efficiency of the stomatal machinery is quite remarkable.

Water potential difference $\Delta\Psi$			Hydraulic conductance L_p m bar^{-1} s^{-1}	Hydraulic conductivity L m^2 bar^{-1} s^{-1}
$\Delta\Psi$ (vap) (bar)	$+\Delta\Psi$ (temp) (bar)	$=\Delta\Psi$ (bar)		
600	400	1,000	10^{-8}	2×10^{-13}
600	400	1,000	10^{-9}	2×10^{-13}
600	400	1,000	10^{-11}	2×10^{-18}
600	400	1,000	10^{-12}	2×10^{-18}

Gaseous conductance *R* and equivalent hydraulic conductance L_p

It is possible to convert resistances or conductances into the equivalent hydraulic conductance L_p provided the conditions are known (or can be estimated) especially the temperature and relative humidity of the gas phase. The simplest way to do this is based on the equivalent hydraulic conductivity L for water vapour which can be deduced from Fig. 6.9.

From Equation [6.3] $J_{wv} \approx D/x \cdot \Delta c$ and from Equation [6.4] $R = x/D$. The diffusive path $x = L/L_p$ (see Equation [1.4]). Combining these equations we find

$$L_p = \frac{L}{DR} \qquad [6.10]$$

Hydraulic conductance (equivalent)	L_p	$L(MLT)^{-1}T^{-1}$	m (bar)$^{-1}$ s^{-1} $\times 10^5$
Hydraulic conductivity (equivalent)	L	$L^2(ML^{-1}T^{-2})^{-1}T^{-1}$	m^2 (bar)$^{-1}$ s^{-1} $\times 10^5$
Diffusion coefficient of water in air	D	L^2T^{-1}	m^2 s^{-1}
Diffusive resistance	R	TL^{-1}	s m^{-1}

Using Equation [6.10] and using values for R of the order given in Table 6.2 we can compute L_p for air and stomatal pores and also cuticle. These values are presented in Table 6.5, assuming the unstirred air layer is 1 mm thick. The stomatal throat is assumed to be 10 μm and 1 per cent of the leaf area. The equivalent hydraulic conductivity, L, of air and stomata is of course the same. The high L value for cuticle supports the calculation in Table 6.4.

For more conventional approaches to these problems see for example Jones (1976).

Table 6.5 Conversion of typical diffusive resistances R to equivalent hydraulic conductances L_p of the gas, stomatal and cuticular phases involved in transpiration. The stomatal path x is assumed to be 10^{-4} m deep and 1 per cent of the superficial area.

Phase	$R\,(\mathrm{s\,m^{-1}})$	$1/R\,(\mathrm{m\,s^{-1}})$	$L\,\mathrm{m\,(bar)^{1}\,s^{-1}}$	x (m)	$L_p\,\mathrm{m\,(bar)^{-1}\,s^{-1}}$
Air	100	10^{-2}	2×10^{13}	10^{-3}	2×10^{-10}
Stomata	1,000	10^{-3}	2×10^{13}	$10^{2}(10^{-4})$	2×10^{-11}
Cuticle	10,000	10^{-4}	2×10^{19}	10^{-7}	2×10^{-12}

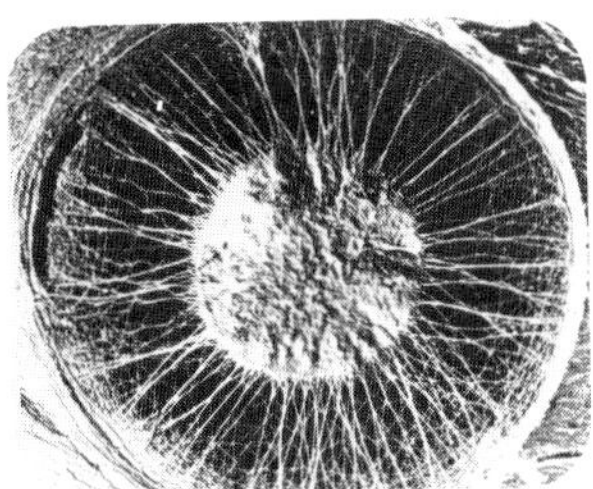

Chapter 7

The control of stomatal aperture

Stomatal control systems have presented one of the most challenging and complex problems in plant physiology. The problem is still not fully resolved, but considerable progress has been made recently. Scientific investigation has proceeded on the lines of Occam's razor, which states that unless there is good reason to the contrary the simplest hypothesis must be tested before considering more complex hypotheses. It is beyond the scope of this book to follow the historical sequence in detail, though it makes a fascinating story. We will consider stomatal structure, measurement of aperture and then separate the two major factors which through their interaction have greatly increased the difficulty of interpreting stomatal control – passive and active mechanisms.

Stomatal apparatus

A stoma is composed of two guard cells which surround a stomatal pore, set in a sheet of epidermal cells (Fig. 7.1) which often interlock like pieces of jigsaw. The cells in the immediate vicinity of a stoma may differ from other epidermal cells, bearing a close relation to guard cells ontogenetically. Generally guard cells of dicotyledons are thickened towards the enclosed pore with the effect that if the cells become less turgid the pore closes, but an increase in turgor produces opening owing to the differential capacity of the cell wall to expand. Stomatal pores may be sculpted so as to produce a series of chambers, some openings of which may be permanently open (as in

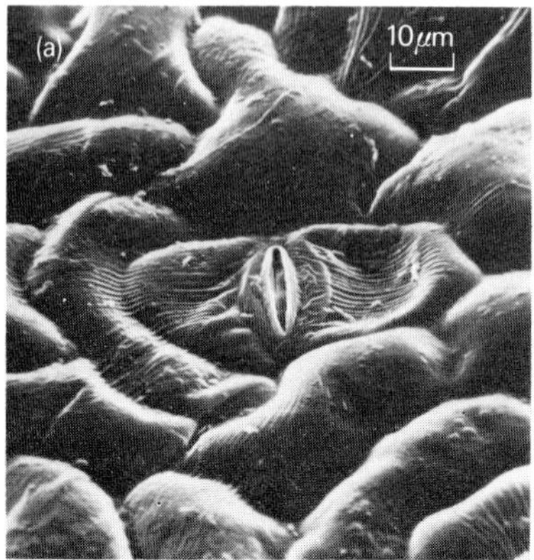

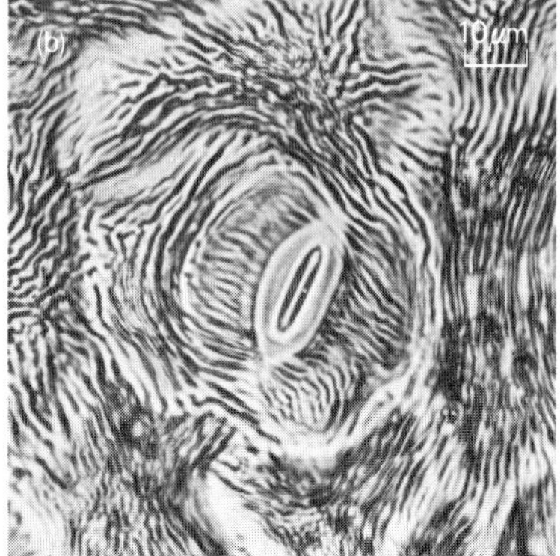

Fig. 7.1 (*a*) Scanning electron microscope image of a *Ricinus* leaf stoma which is typical of many dicotyledons. Note the permanent outer-lips bordering the pore – the throat of the pore constitutes the variable turgor mechanism. The 'stretch' lines appear to be permanent features of the cuticle which run over many epidermal cells (Milburn and Sprent, unpublished). (*b*) Comparable silicone rubber imprint (see App. 16).

the case of *Ruscus aculeatus* leaves) or filled with hairs (as in the *Ericaceae*). In the floating leaves of hydrophytes all the stomata are uppermost, in trees they tend to be on the lower epidermis, but in herbaceous plants stomata are found in considerable numbers on both upper and lower epidermes. In Table 7.1 some typical values are

Table 7.1 Typical stomatal frequencies, pore lengths and pore areas as percentages of total leaf areas (from Meidner and Mansfield, 1968)

Plant		**Frequencies per mm^2**		**Length of pore μm**		**Pore area % total leaf area; (pore width 6 μm)**
		upper	lower	upper	lower	
Herbs	*Allium cepa*	175	175	24	24	2.0
Herbs	*Avena sativa*	50	45	20	19	0.5
Herbs	*Triticum vulgare*	50	40	28	28	0.63
Herbs	*Zea mays*	98	108	12	16	0.7
Herbs	*Helianthus annuus*	120	175	15	17	1.1
Herbs	*Pelargonium zonale*	29	179	24	23	1.2
Herbs	*Ricinus communis*	182	270	12	24	2.1
Herbs	*Vicia faba*	65	75	28	28	1.0
Trees	*Pinus sylvestris*	120	120	20	20	1.2
Trees	*Quercus robur*	0	340	0	10	0.8
Trees	*Tilia europea*	0	370	0	10	0.9

listed obtained from microscopic measurements. These may be made on fresh leaves, cleared leaves, epidermal strips or leaf surface imprints made using, for example, silicone rubbers. Stomatal frequencies and sizes vary depending on the growth of the plants concerned; large leaves tend to have larger stomata which are more widely spaced than their smaller counterparts when the same species has produced small leaves in response to drought.

The conductance of stomatal pores

It is possible to deduce stomatal diffusive conductance and resistance from a theoretical consideration of the mean dimensions of the pores. The subject is technically complex for several reasons not least of which is the definition of a 'stomatal pore'. According to one view the pore is the stomatal throat itself; others, dating from the work of Brown and Escombe (1900), have added 'end corrections', taking into account the diffusive spread of vapour at each end of the throat. There is now strong evidence that considerable evaporation takes place in the throat itself affecting an inner 'end correction' – the outer 'end correction' is strongly subject to air flow (wind) since it can be regarded as part of the unstirred gas layer. In addition to these problems stomatal pores can be oval, diamond or rectangular in cross-section with complex additional architecture such as cuticular ridges and hairs.

An equation was derived for cylindrical tubes by Brown and Escombe which has been confirmed subsequently for stomatal gaseous resistance R_s. The expression Πd is the 'end correction' and Πd^2 represents the cross-sectional area of each pore.

$$R_s = \frac{4x}{ND} + \frac{\Pi d}{\Pi d^2} \qquad [7.1]$$

where d is the diameter of the cylindrical tube and other units are as below. In reality stomatal pores are more nearly rectangular or diamond shaped in cross-section, so the theory can be modified to the simple form below which ignores the end correction and assumes the area is that of a diamond-cross-section pore. Several more complex equations have been proposed.

$$R_s = x/0.5\,ab\,ND \qquad [7.2]$$

Diffusive resistance (stomata)	R_s	TL^{-1}	$\mathrm{s\,m^{-1}}$ (in still air)
Depth of stomatal pore (mean)	x	L	m
Stomatal width (mean)	a	L	m
Stomatal length (mean)	b	L	m
Stomatal frequency	N	L^{-2}	$\mathrm{m^{-2}}$
Diffusion coefficient H_2O in air	D	L^2T^{-1}	$\mathrm{m^2\,s^{-1}}$ (at given °C)
	Π	–	3.141

Taking typical figures for the lower epidermis of a *Ricinus* leaf at 20°C where x is 24 μm, a is 6 μm and b is 38 μm and N is $2.7 \times 10^8\,m^{-2}$, R_s is approximately $31\,s\,m^{-1}$. If the pore closes so that $a = 0.25\,\mu m$ then R_s increases to $751\,s\,m^{-1}$.

This value ignores the effect of the unstirred gas layer which in still air would exert a greater hindrance to diffusion than the stomatal apparatus itself. In wind the stomata would tend to close.

Published values for R_s vary with the species and method of determination. Data from Meidner and Mansfield (1968) and Monteith (1973) indicate R_s commonly falls to around $150\,s\,m^{-1}$ in mesophytes but rises to about $3{,}000\,s\,m^{-1}$ in conifers and xerophytes which have sunken stomata. A realistic approach might be to introduce a correction factor for the species chosen.

It is possible to convert diffusive resistances into diffusive conductances, by taking the reciprocal, or into equivalent hydraulic conductances using Equation [6.10]. At 20°C and 80 per cent mean r.h. L_p for a stoma with $R_s = 31\,s\,m^{-1}$ gives an $L_p = 3.2 \times 10^{-11}$ m (bar)$^{-1}$: if R_s is $751\,s\,m^{-1}$, $L_p = 1.3 \times 10^{-10}\,m\,(bar)^{-1}\,s^{-1}$

The measurement of leaf porosity

Individual measurements of stomatal apertures are tedious to make because larger numbers are necessary per sample to monitor the whole population, and even when this has been done it is not easy to use the measurements to deduce porosity. It is easier to monitor porosity using a porometer. In principle the cup of a porometer should cover a large population sample. In *viscous flow porometers* air is drawn through the leaf in various ways. The rate of gas flow measures the porosity of the leaf, which varies according to stomatal behaviour. Of the many porometer designs, the porometer in Fig. 7.2 is a variant of Knight's porometer which has proved reliable for classwork and is reasonably foolproof. Air is drawn into the cup through the leaf by a gas suction device consisting of a constant-pressure aspirator system, so that the suction generated corresponds to the height of the jet above the siphon efflux level. The rate of bubbling may be timed directly or this process may be automated using a microphone and tape recorder (see inset) with a scaler. For accurate work, the porometer should be calibrated, because the bubbles increase slightly in volume when flow-rates are high and over long periods attention must be paid to changes in barometric pressure which can affect the enclosed gas volume.

Another type of porometer is the *diffusion porometer* (Fig. 7.3), which measures the tendency of a leaf surface to lose water. Various

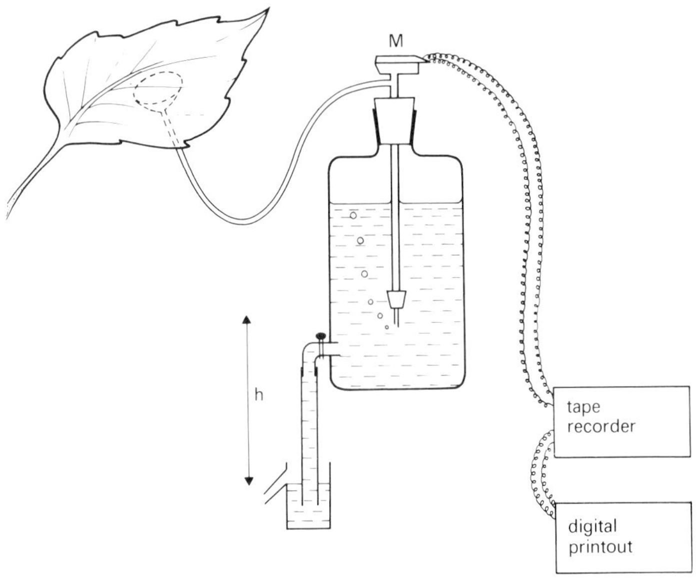

Fig. 7.2 Viscous flow porometer – modified version of Knight's porometer based on a constant suction aspirator. The water siphon arrangement h draws a flow of air to the leaf which can be counted as bubbles either manually or automatically using a microphone transducer system M.

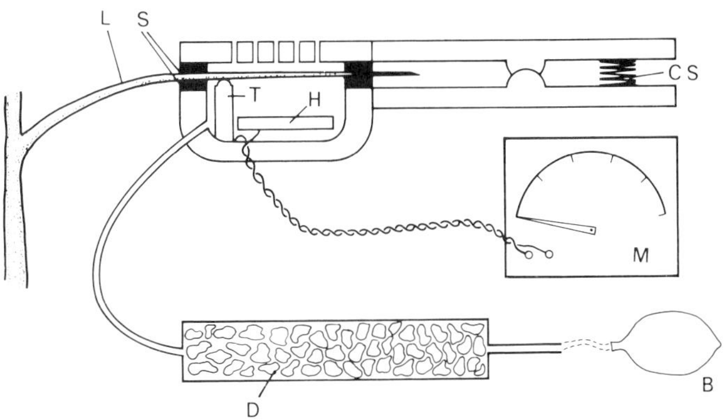

Fig. 7.3 Diffusion porometer showing cup held by clip spring CS sealed (S) on leaf L. Humidity change is registered on meter M by hydrometric sensor H as dry air from D is moistened by transpiration. The bulb B serves to replace dry air, and thermistor T measures the leaf temperature.

systems have been used to do this in the past. Papers impregnated with cobalt chloride, when attached to a leaf surface, changed in colour at a rate proportional to the absorption of water. Darwin used a horn hygroscope, measuring the curvature of a sliver of hygroscopic horn held against a leaf surface. Most modern procedure is based on electronic moisture-sensor systems. In principle the porometer cup, filled with dry air, is pressed on a leaf surface and the time is taken for a standard moisture reading to register. The porometer can be calibrated against filter paper moistened with water or a range of osmotic solutions so that the effective water potential of the leaf surface can be measured. This in turn can be related to the flux of water vapour from the leaf, so providing simultaneously a measure of the transpiration rate and also the diffusion capability of the leaf. The problem with diffusion porometry is that the stomata can change aperture quite rapidly when their environment changes (as caused by covering them or subjecting them to air which is dry or deficient in carbon dioxide). Also the energy balance of the leaf, on which evaporation depends, is altered by a cup and a thermistor is often included in the cup design to monitor leaf temperatures.

'Passive' mechanical control

There is now little doubt, though the idea was at one time challenged, that stomata operate so as to protect a plant from the effects of desiccation. According to this view a restriction of water supply should induce stomata to adopt their protective function and close. Enigmatically, however, stomata, monitored by porometry, are found to open widely if a leaf is excised. Though the stomata eventually close, the opening phase is far too significant to be accounted for by experimental error. The porometer airflow may reach twice the pre-excision value (PEV) and it often remains above PEV for 20 minutes.

Two hypotheses were advanced to explain the observation: (1) Darwin and Pertz (1911) suggested that the opening was caused by the shrinkage, through loss of turgor, of the epidermal cells, which normally tend to compress the turgid guard cells. If the epidermal cells lost water more rapidly than guard cells, the stomatal pores would expand passively under the released tissue pressure of the guard cells; (2) Ivanoff (1928) explained the observation quite differently by suggesting that when a leaf is excised water will be pulled back from the severed xylem conduits and displaced by air. This redistribution of water raises the water balance of guard cells increasing their turgor pressure so opening the stomatal pore more widely.

Neither hypothesis required any physiological change on the part of the guard cells; the effect was agreed to be a 'passive' mechanism. The first hypothesis implied that stomatal control was severely limited, because it was not able to cope with sudden changes in water balance. The second hypothesis implied that the quantity of water released from severed xylem was sufficient to account for turgor increase, despite continuing transpiration in the absence of a water supply. Heath (1938) had shown that if epidermal cells were punctured surrounding a stoma the pore tended to open favouring hypothesis (1). Willis, Yemm and Balsubramanian (1963) supported hypothesis (1), having shown that opening was far greater when the transpiring leaves were very turgid than when their water balance was reduced. In the 1960s the controversy again flared. Among the leading protagonists for hypothesis (1) was Meidner, who showed that the volume of water which might be released from the xylem was probably insignificant in comparison with the continuing transpiration rate of an excised leaf, but this quantity was dependent on conduit volumes, which may vary. Unfortunately none of these observations gave a sufficiently clear demonstration to constitute proof.

The solution lay in the fact that hypothesis (1) explained stomatal opening as a response to a rapid *decrease* in water potential Ψ_p, whereas hypothesis (2) attributed opening to an *increase* in water potential Ψ_p. If it could be shown that (A) a decrease in Ψ_p produced opening or alternatively that (B) an increase in Ψ_p induced closure, this would effectively clinch the argument in favour of hypothesis (1); the converse would favour hypothesis (2). Evidence for alternative (A) was provided by Raschke (1970), who observed that if a leaf stalk were placed in a vacuum flask, and the stomatal aperture monitored as vacuum was switched on, the aperture increased. Additional evidence for hypothesis (1) based on alternative (B) had been obtained independently. By applying positive pressure to the water supply of a *Ricinus* leaf stomatal aperture was reduced though it increased again when the pressure was removed (Fig. 7.4). The evidence convincingly favours the Darwin and Pertz hypothesis. The effect is transitory, tending to disappear as epidermal cells and guard cells equilibrate with each other.

Clearly when stomatal changes are investigated many factors (e.g. radiation, humidity, water potential) can influence aperture. It is necessary to ascertain whether observed effects are likely to be passive, through the mechanical intervention of epidermal cell turgor, or direct effects on guard cell physiology. Another way in which passive opening may produce complex effects is through cycling, which gives interesting information regarding the speed with which stomata may respond passively.

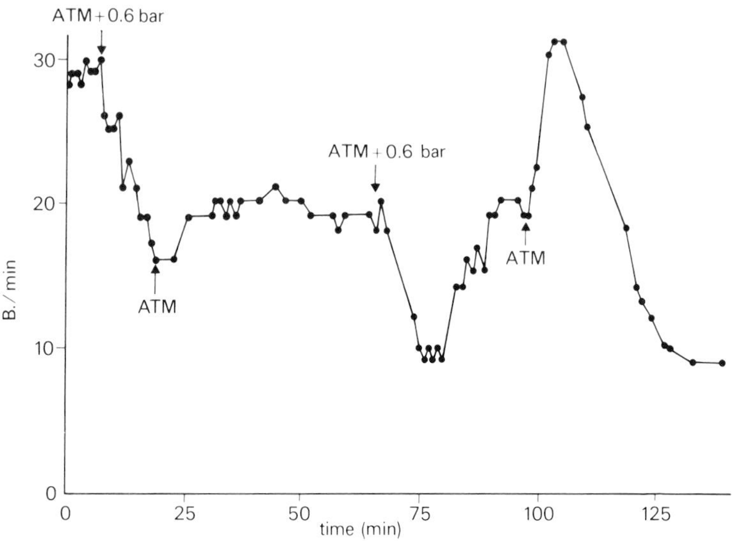

Fig. 7.4 Effect of pressurising the water supply to a *Ricinus* leaf monitored by a fixed-cup Knight's porometer. After an increase in pressure *P* to 0.6 bar above atmospheric pressure *A* there is a transitory decrease in stomatal aperture caused by an imbalance in epidermal–stomatal turgor pressures. The effect is reversed when the increased pressure is removed. Porosity is indicated on a relative scale of bubbles per minute (B/min) (Milburn, unpublished).

Midday closure

An intriguing example of the complexity of environmental changes is the midday closure of stomata, first observed by Loftfield (1921). At first he doubted his observation, with justification, because closure occurs when illumination and so stomatal apertures should be greatest. Furthermore the phenomenon was apparent in some plants but not others; it is frequently encountered in crop plants such as sugarcane (*Saccharum officinarum*) and even the oil palm (*Elaeis guiniensis*, see Rees, 1961).

The mechanism is still puzzling. On one hand closure might be caused by a local water deficiency induced by excessive transpiration. We know deficits can induce endogenous formation of abscisic acid which causes closure. There is evidence, however, that irradiation causes a rise in temperature which increases respiration more than photosynthetic assimilation. Thus closure is the result of high concentrations of carbon dioxide. This view is supported by the fact that if stomata are flushed with carbon-dioxide-free air the stomata can be induced to reopen (Meidner and Heath, 1959).

Stomatal cycling
When a leaf has been darkened stomata normally close, but on reillumination they open widely, indeed they may do this to such an extent that they 'overshoot'. In a constant environment they exhibit a series of opening and closing responses while they 'hunt' for a new dynamic equilibrium between opening and closure responses. Eventually such an oscillation tends to disappear, but in certain circumstances this may persist.

Many similar experiments in which rapid changes in water potential are induced, for example, by root cooling (see Ch. 4) or resuscitation after wilting, give rise to this response. The implications are that the stomata are induced to open and do so to such an extent that the leaf suffers partial desiccation which induces closure, again raising the water balance of the guard cells and inducing opening. Each stoma acts independently with a slightly different rate of response. Consequently after a sudden change all coincide in phase. Thus violent oscillations in net aperture are detectable. Gradually synchrony is lost once more producing a semblance of equilibrium. The interesting implication is that, when we attempt to measure the mean stomatal aperture of a leaf, not only are we dealing with a large and varied population but each individual stoma is cycling continuously.

The study of cycling is complex, because it can be caused by both passive and active components. Systems with capacitance and a conductivity can be arranged so as to oscillate, and indeed this is the principle of tuning circuits in electronics. Increasingly attempts are being made to resolve cycling by making electronic models incorporating resistance and capacitors to represent the complexities of plant anatomy (see e.g. Cowan, 1972).

Active mechanisms

An early observation on stomatal physiology was the well-established fact that illumination tends to induce opening while darkening produces closure. Von Mohl (1956) proposed that guard cell turgor changed because sugars were produced photosynthetically in guard cells. Several objections invalidate this hypothesis. For example, stomata open more rapidly than could be accounted for by photosynthetic assimilation. Furthermore, stomata open widely in air from which carbon dioxide is removed, which should be expected to prevent photosynthesis and so opening (Lloyd, 1908).

Attention was then focused on the possibility that light might mediate in the digestion and synthesis of starch which was present in guard cells and which tended to appear and disappear on a diurnal basis by enzymic action. Attempts were made to explain these changes in response to changes in acidity, since Sayre (1946) had observed that stomatal opening was correlated with increased pH of the guard cells, an observation later established by other workers. A favoured concept was that pH might cause a shift in substrates through an influence on enzymic processes, a view which gained support when it was discovered that the hydrolysis of starch to produce glucose phosphate was correlated with an increase in pH.

$$\text{starch} + \text{inorganic } PO_4^- \underset{pH5}{\overset{pH7}{\rightleftharpoons}} \text{phosphorylase glucose-1-}PO_4$$

Objections were immediately apparent because if inorganic phosphate were unbound there would be no difference in osmotic potential from glucose-1-phosphate synthesis. Furthermore phosphorylase has not been detected in guard cells. A further problem was caused by an observation by Heath (1949) that though starch content varied diurnally, if the normal sequence of illumination or darkness were interrupted, the stomata responded but starch content did not obviously change. One observation favouring a starch-digestion hypothesis was that wilting of a leaf led to the rapid accumulation of starch (Iljin, 1957), now an established observation. Meantime autoradiography showed clearly that guard cells could assimilate $^{14}CO_2$ photosynthetically to produce starch, but as already pointed out such a photosynthetic mechanism would not explain stomatal opening in carbon-dioxide-free air.

A key to much modern work has stemmed from observations by Iljin (1957) and Arends (1926) that stomatal opening and the disappearance of stomatal starch can be controlled reversibly by adjusting the cation balance of isolated stomata. Monovalent solutions, especially potassium, produced wide opening. Divalent cations (Ca^{2+}, Mg^{2+}) countered this effect. The interpretation of stomatal interactions has been greatly simplified by working on stomata either in epidermal strips or in complete isolation.

Recent developments to elucidate stomatal control

In recent years it has become evident that the accumulation of potassium by guard cells, from sap in cell walls adjacent to stomata, is strongly correlated with stomatal opening. This has been shown by chemical tests (cobaltinitrate stain), by electron microprobe analysis, and Penny and Bowling (1974) were also able to show, using K^+

sensitive microelectrodes inserted in *Commelina* epidermal cells, that K^+ accumulated (i.e. was lost from guard cells) as closure occurred. It is uncertain if the potassium is itself actively transported, since it could diffuse from pumping of organic acids or, even more likely, from proton pumping from guard cells under the agency of metabolic activity. At a stroke the proton secretion hypothesis offers a mechanism to modify the osmotic potential of guard cells. It also offers an explanation for the effect of inhibitors (cyanide, azide and dinitrophenol) in producing stomatal closure. Such a mechanism could also operate to produce stomatal opening in carbon-dioxide-free air.

Progress has been made to remove the effect of other tissues on stomata. It could be argued that this removal of normal cell associations produces an artificial situation, but evidence has accumulated to the contrary. Guard cells on detached starved leaves retain starch long after it has disappeared in other cells suggesting that guard cells are isolated. Very few plasmodesmata have been found between guard cells and adjacent cells, further supporting this view. There is no doubt that guard cells can operate apparently normally when the less robust epidermal cells have been removed. Squire and Mansfield (1972) have used an acid medium (pH 5) and they (and others) have used a rolling technique to kill epidermal, but not guard, cells in *Commelina*.

Edwards and Meidner (1975) used a pressure-transducer microprobe to measure guard cell turgor pressures directly. Somewhat surprisingly the pressures are smaller than might have been expected if the guard cells approached full turgor. (Even abnormally wide apertures could be closed by pressurising subsidiary cells 4 to 7 bar.) The implication is that guard cells, on account of their exposed position, normally operate at a water potential considerably lower than most epidermal cells.

The finding that stomata respond to humidity of the surrounding air rather than the water potential of tissues certainly indicates an effect of water potential. However, the assumption that xylem water potentials are *closely* correlated with stomatal water potentials is not always valid because of the tenuous water conduction pathway separating and almost isolating guard cells from vascular tissues.

It has been known for many years that guard cells behave rhythmically. The significance and persistence of such circadian rhythms have been investigated by Martin and Meidner (1972), who have shown that this rhythm can induce a certain amount of stomatal opening even in total darkness.

Many questions still await resolution. For example, is the fact that stomata respond to blue light an indication of a photoactive pumping mechanism? Does high temperature, which induces stomatal closure, do so by raising carbon-dioxide concentrations to toxic levels?

The effects of chemicals on guard cell turgor

Many inhibitors tend to induce stomatal closure. Perhaps this is not surprising because any interference with metabolism might induce a loss of guard cell turgor. Remarkably, however, certain chemicals have the opposite effect. It was first observed that almond plants suffering from the wilt-inducing pathogen *Fusicoccum* transpired much faster than uninfected control plants. This led to the isolation of the active principle, a fungal toxin called fusicoccin which induces wilting even when applied to uninfected plant material. Fusicoccin at a concentration of $10\,\mu M$ stimulates stomatal opening, even in darkness, which would explain wilting through loss of stomatal control (Squire and Mansfield, 1974). Fusicoccin also enhances the accumulation of potassium in guard cells. Presumably the changes in guard cell turgor occur indirectly, through modification of active accumulation of ions. In these experiments a fungal toxin has been used as a tool to study guard cell metabolism.

An even more interesting discovery is that abscisic acid (ABA), which despite its name seems to have little natural role in promoting abscission, has a very marked effect on stomata (Wright, 1969). It is well established that ABA is produced endogenously in plant tissue subjected to a water deficit, that ABA induces stomatal closure in concentrations of $10^{-5}M$, and there is a detectable effect at $10^{-10}M$ (Mansfield, 1976). This release of ABA seems to explain why transpiration is severely checked after plants have been given a single severe water deficit (Fig. 7.5). The metabolism of ABA seems to be involved in the gradual recovery process which takes place over several days (see Meidner and Mansfield, 1968).

What is fascinating about the action of fusicoccin and ABA is the low concentrations required to produce an effect. Also ABA seems to exert a hormonal control of stomata naturally in the intact plant. Recently evidence indicates that all *trans*-farnesol, which like ABA is a sesquiterpenol and which accumulates in water stressed sorghum can also induce stomatal closure (Mansfield, 1976). The present situation is strongly reminiscent of the complex sequence of hormonal interactions producing homeostatic control of ion- and water-balance in animals, involving angiotensins, vasopressin and antidiuretic hormones (see Roddie, 1971).

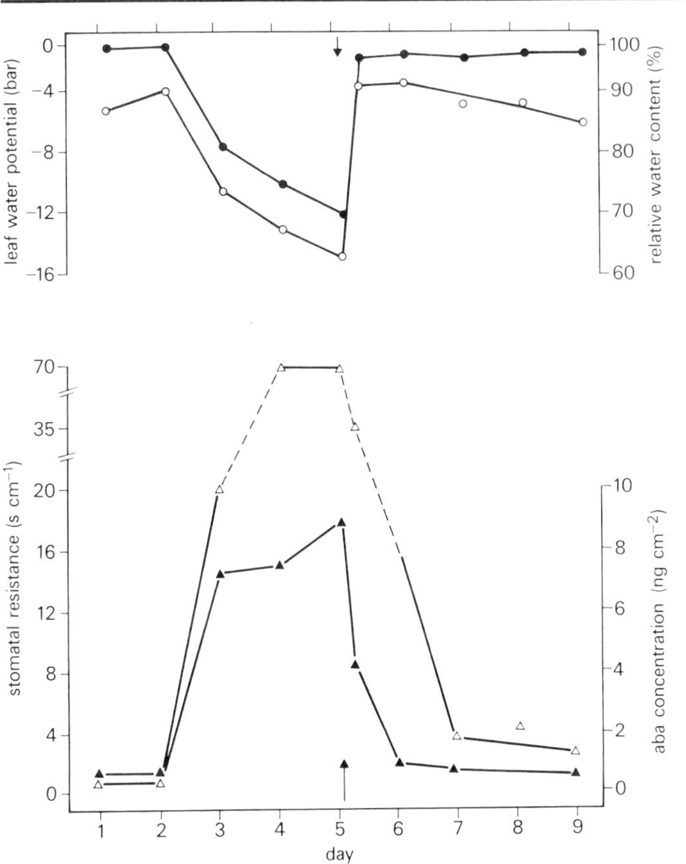

Fig. 7.5 Changes in *Zea mays* leaf water potential (○), relative water content (●) ABA(▲) and diffusive resistance of the stomata (△) during a stress–recovery experiment. The arrows indicate rewatering. Note that the reduction of diffusive resistance lagged about a day behind a reduction in ABA level. (From Beardsell and Cohen, 1974.)

Summary of factors controlling stomatal aperture

Environmental factors

Light tends to cause stomatal opening, and high temperature, below a lethal threshold, also favours this tendency. Relative humidity of the air influences guard cell turgor directly, because stomata are subject to rapid loss of water depending on the water potential gradient.

Table 7.2 Factors influencing the guard cell environment which promote stomatal opening. When reversed these treatments promote closure. CAM plants can be exceptions to these general rules (see Ch. 8).

Environmental factor acting on closed stomata		Changes in factor promoting stomatal opening	Typical values encompassed normally
Light (visible radiation)	→	Increased	0 to 1,000 W m^{-2}
Temperature (and i.r. radiation)	→	Increased*	20 to 40°C
Water potential Ψ of leaf	→	Increased	e.g. −7 to −5 bar)
Humidity of air	→	Increased	30 to 95% r.h.
Fungal toxins (e.g. fusicoccin)	→	Increased	0–10 μM
Carbon dioxide concentration	→	Reduced	0.033 to 0.01% by vol.
Tissue pressure (epidermal cells)	→	Reduced	A few bar
ABA concentration in sap	→	Reduced	e.g. 10^{-3} M to below 10^{-8} M
Atmospheric pollutants (e.g. SO_2)	→	Reduced	0 to 2.5×10^{-5}% by vol.
Metabolic inhibitors (e.g. DPN)	→	Reduced	–

* Temperature decreases stomatal opening if excessive.

Wind can aggravate water potential gradients further. The summarised list (Table 7.2) showing the effect of external factors on stomatal aperture must be applied cautiously because many factors interact. Though it is not easy to predict the outcome of such interactions, adverse effects usually promote closure thus low carbon dioxide concentrations induce abnormally wide stomatal apertures, but pollutants such as sulphur dioxide have the reverse effect.

Internal factors

Changes in water potential of a plant have complex effects on stomatal aperture and can cause temporary passive opening or stomatal closure. Hormones such as ABA can induce stomatal closure and apparently perform this role in nature by preventing potassium accumulation in guard cells. Ionic balance and carbohydrate reserves both exert metabolic control of stomatal aperture, but the main driving mechanism is probably mediated by proton extrusion pumps which also induce pH changes in guard cells. Light and temperature probably enhance extrusion from guard cells and fungal toxins such as fusicoccin may play a similar role. The main internal factors are listed in Table 7.3.

Table 7.3 Reversible changes in guard cell physiology associated with stomatal opening

Internal factors when stomata are closed		**Changes in factor promoting opening**
Proton balance around pH 5	→	Increased to around pH 6
Cation balance low, especially K^+	→	Increased
Anion balance low, especially Cl^- and malate	→	Increased
Sugar concentration low	→	Increased
Turgor pressure low, around 5 bar	→	Increased a few bar (e.g. 5 to 8 bar)
Osmotic potential ranges -6 to -25 bar	→	Increased a few bar (e.g. -6 to -4 bar)
ATP concentration high (with ATPase)	→	Decreased as ADP is formed
Starch deposition high	→	Decreased

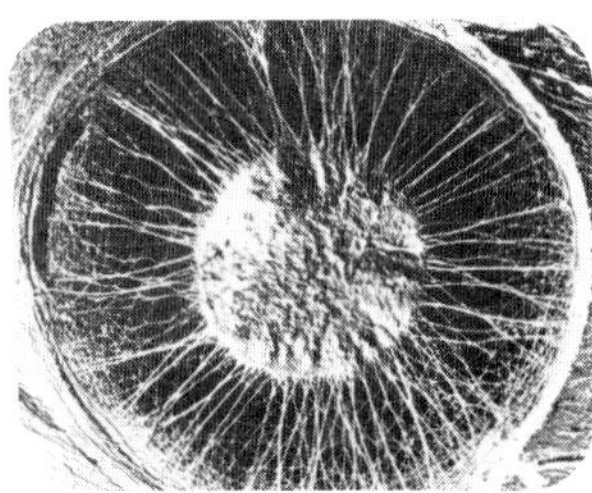

Chapter 8

Water and growth

Introduction

Plants obviously require water for growth because water is essential for cell volume increase by vacuolation which is responsible for growth. During drought a restricted water supply first reduces the growth rate, eventually stopping it altogether. In the normal situation a plant must carefully regulate its expansion because the uptake of water must compensate for cell enlargement which may simultaneously provide plant support. It must be admitted that the role of growth in plant–water relations is not yet fully understood but the progress made is a fascinating story.

Early submerged leaf experiments

In July 1724 Stephen Hales cut off an apple tree branch and submerged it in water while measuring the uptake of water by the cut stump. The water bath prevented water loss by transpiration, but was not available for uptake because the leaf surface is waterproofed, consequently the woody stump continued to absorb water for three days. The shoot seemed unharmed by its immersion and the stump absorbed water vigorously when the shoot was exposed to air once more.

Similar experiments were performed by Dixon before 1900. He too observed persistent uptake by submerged shoots and leaves. Even after several days dyes were carried into the submerged shoots leading Dixon, the great elucidator of the physical basis of sap ascent, to postulate this uptake was driven by a mesophyll pumping system which caused 'subaqueous transpiration'. Later, when it was suggested that his observations were mistaken and caused merely by a persistent water deficit, Dixon and Barlee (1940) reinvestigated the observation, reconfirmed its existence and showed that water uptake was suppressed by low temperature or restrictions in oxygen supply which supported the hypothesis that mesophyll cells pumped water using metabolic energy.

Complications in measuring water deficits

Quite different experiments were designed to establish a satisfactory basis for measuring plant water deficits. Stocker (1929) and later Weatherley (1950) proposed that a suitable reference point would be the water content of a tissue when it had imbibed water to repletion. A correction was applied to counteract possible errors caused by possible variation in dry matter content (see Ch. 3).

Unfortunately, contrasting with theoretical expectations, leaf discs floated on water continued to absorb water long after the deficit uptake seemed to be complete (Fig. 8.1). There was no possibility that this prolonged uptake was caused by dry weight increase from photosynthesis because the discs were floated under just sufficient

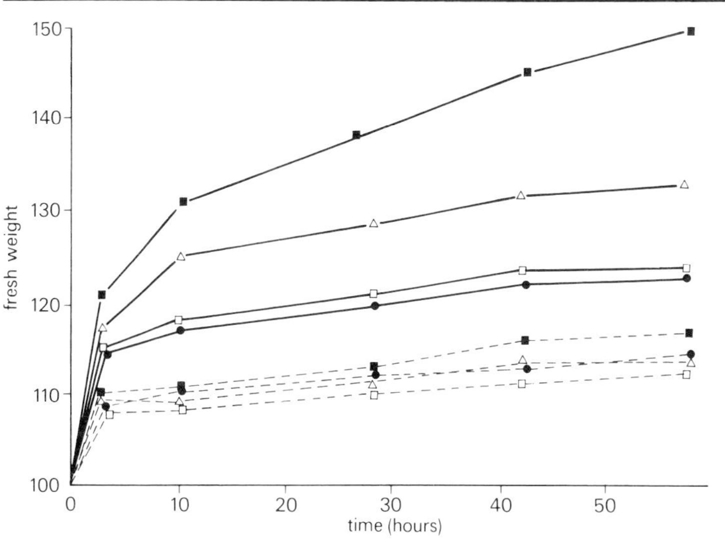

Fig. 8.1 The uptake of water by *Ricinus* leaf discs floated on water at 32°C (upper curves) or at 1°C (lower dashed curves) against time. Younger leaf 1(■) absorbs proportionally more warm water for expansion than leaves of increasing age 2(△), 3 (□) and 4 (●) but the corresponding uptake at 1°C was similar for all ages (from Milburn and Weatherley, 1971).

illumination to compensate for respiratory losses of dry matter (i.e. at the compensation point). Since growth seemed responsible for prolonged uptake Barrs and Weatherley (1962) tried to suppress it completely by floating discs on cold water at 1°C. Obviously if growth could be suppressed completely, deficit uptake could be measured at any time because growth increments would be eliminated. Enigmatically cold water suppressed not only the slow prolonged uptake but also a fraction of the rapid uptake (presumably deficit) phase also. Suspecting that cold conditions reduced uptake of water by protoplasts, in much the same way that cold reduces water uptake by roots, Barrs and Weatherley (1962) advised that deficits would best be measured at ambient temperatures allowing 4 hours for completion of the rapid uptake phase. This method has been widely adopted, but is open to objections.

Further submerged leaf and shoot experiments

Weatherley (1963) described ingenious experiments in which he attempted to partition leaf water deficits into two components based on a dynamic analysis. On this view water flow into cell walls should be rapid but uptake into protoplasts should be slower on account of the hydraulic resistance of the plasmalemma and tonoplast membranes. He tested this hypothesis by submerging transpiring leaves in water (to stop transpiration) and then measured the rate of uptake of water. Results were gratifying: two phases could be distinguished and furthermore each phase seemed to be logarithmic (i.e. log plots were linear) suggesting the anticipated two superimposed phases of the leaf water deficit, i.e. cell walls and protoplast or vacuolar components. Furthermore when leaves were plunged into cold water only the rapid phase was detectable but if warm water was substituted as a bathing medium the slower 'vacuolar' phase could be measured. These results seemed to confirm Barrs's and Weatherley's (1962) experiments and supported the view that cold conditions partitioned leaf water deficits into wall and protoplast components.

On the basis of these experiments the rapid, cold-insensitive uptake should correspond with cell-wall water deficits. However Milburn and Weatherley (see Milburn, 1964) found that in *Ricinus* the cell wall total volume could occupy 15 per cent of the leaf cell volume. Nevertheless cold-insensitive uptake could account for about 50 per cent of the leaf water content and during this uptake leaves recovered from wilting, becoming stiffly turgid. These results suggested that cold uptake was in fact the true wall-plus-protoplast deficit. This view was supported by the finding that the warm-uptake phase was greater the earlier the stage of leaf development, suggesting a correlation with growth (Fig. 8.1). Furthermore the rapid cold-sensitive uptake of water could be enhanced by withholding water from plants, a treatment known to develop water stress and check growth. On floating discs from water-stressed leaves they were found to increase in area by a staggering 20% in 5 hours, but much less if floated on cold water (Milburn and Weatherley, 1971). All the latter experiments meant that growth played a much larger role in both rapid and prolonged water uptake than previously expected. The corollary drawn was that water deficits must be measured near 0°C to reduce complications caused by rapid and also prolonged growth and there is no reason to question this conclusion.

'Subaqueous transpiration' explained

The work described above seemed to indicate that Dixon's and Barlee's experiments could now be explained quite simply. Prolonged uptake of water was caused by growth and not as a result of mesophyll pumps which somehow expelled water from submerged leaves. How else could one explain the continued weight increase in for example floating leaf discs? This supposes the very reasonable hypothesis that the suppressing action of anaerobic conditions on prolonged water uptake inhibited growth.

These hypotheses were tested and proved by Potter and Milburn (1970). Exactly the same behaviour was observed in test plants (e.g. *Chrysanthemum* and *Salix*) as observed earlier in *Ricinus* – young leaves absorbed water more persistently and more rapidly than older leaves. There was no evidence of errors in making measurements, e.g. from water injection of the gas spaces in leaves. A final experiment seemed to clinch the issue conclusively. When shoots were submerged in liquid paraffin oil instead of water for several days they continued to absorb water but none of this could be detected on the leaf surface submerged in oil. Furthermore if the shoots were inverted in the paraffin they tended to revert to a normal position by growth showing few ill-effects from the oil on normal metabolic processes.

The nature of the growth process

In Chapter 3 we considered the water potential of a typical plant cell as having pressure (Ψ_p) and solute (Ψ_s) components. If a cell were not growing these components should reach equilibrium very rapidly. But some imbalance must be caused if growth continues. Undoubtedly the wall area increases and if the cell wall were to expand rapidly a reduction in cell turgor pressure might be measurable. The tendency for water to enter the cell is driven by the water potential gradient, effectively the difference between internal and external water potentials. Solutes within the cell lower the water potential but as the cell volume increases during growth the solutes in the cell are diluted and the internal water potential tends to increase, reducing the water potential gradient. Clearly, cell growth could be controlled either by wall relaxation or alternatively by an increase in solute concentration so increasing the water potential gradient. If the wall relaxes we might expect to measure a *reduction* in turgor pressure of the cell to initiate growth. An increase in solute concentration would induce a transitory *increase* in turgor pressure, soon removed as the wall became stretched in response with a concomitant dilution of solutes through water influx. These alternatives can be expressed diagrammatically in Fig. 8.2. Clearly both processes might occur together, but which is the more significant process quantitatively?

Burstrom (1948) suggested that wall relaxation was the key process

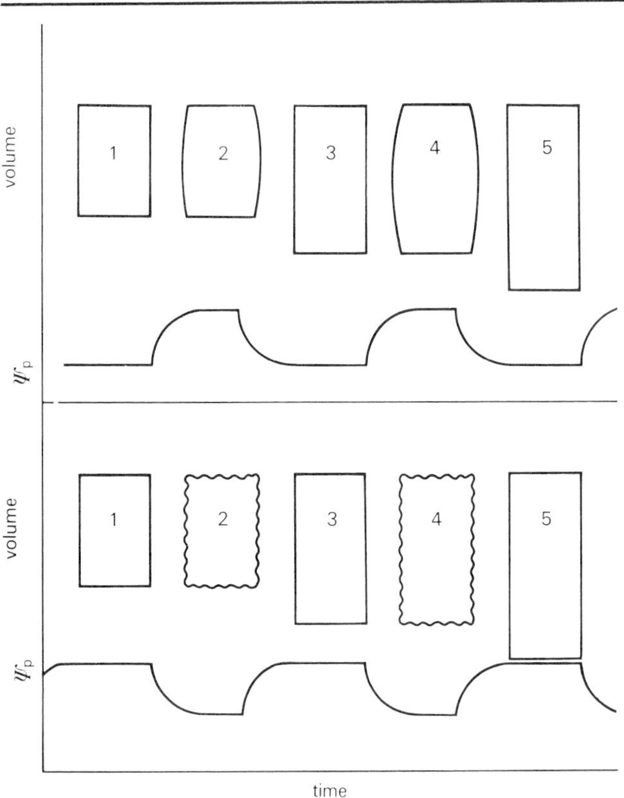

Fig. 8.2 Diagram showing how Ψ_p might vary if cell expansion were dependent on increased pressure (upper curve) which stretched the wall, or alternatively if the wall expanded before pressurisation. Inset shows how expanding cells might appear at stages 1–5 corresponding to the changes in Ψ_p.

in cell growth and evidence in favour of this view has accumulated. But his suggestion led to a fascinating exchange of papers, Thoday, Haines and Weatherley suggesting that if wall relaxation preceded an increase in turgor this must violate Newton's first law of motion which states that every action and reaction must be equal and opposite. The problem was explained by Spanner (1952) who pointed out that in reality a pressure wave would be initiated by wall relaxation followed by rapid adjustment in turgor pressure (or vice versa) over a short but finite time. Indeed if this were not so *any* motion would be impossible.

Various approaches have been made to study growth extension of cells. Green *et al.* (1971) studied the growth rate of filamentous algal cells of *Nitella* by cine-filming them at 3 second intervals. Internal pressures were measured on a bubble-compression gauge inserted into the cell under observation (Fig. 8.3). The following formula was proposed:

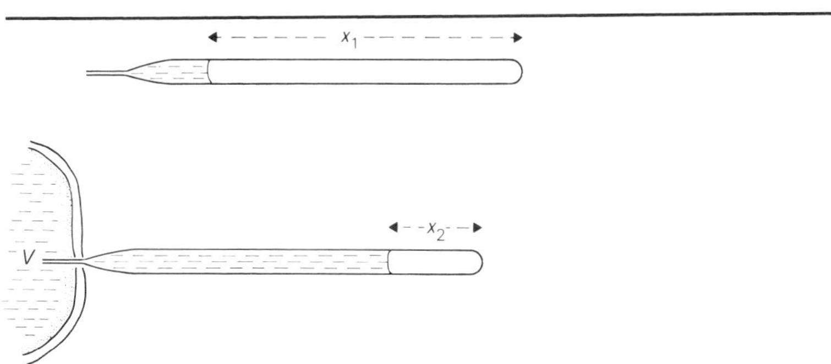

Fig. 8.3 A bubble manometer which can be used to measure fluid pressures filled with water at atmospheric pressure (*above*) then inserted in a plant cell (*below*). The bubble volume, measured by x decreases depending on the pressure. Cell pressure $P_2 = x_1/x_2\, P_1$ where P_1 is atmospheric pressure (1.013 bar).

$$\text{Growth rate} = m(\Psi_p - Y) \qquad (\text{bar h}^{-1}) \qquad [8.1]$$

m represents the rate at which turgor pressure Ψ_p can be maintained as the wall becomes fluid. Since several workers had observed that cells did not necessarily expand even when slightly turgid it could not be assumed that the growth rate was directly proportional to the turgor pressure. Hence the term Y, the yield point, which is the threshold pressure below which growth is zero. Growth of *Nitella* cells could be stopped rather easily by small changes in external Ψ_s. For *Nitella* cells growing in length at about 2 per cent per hour Ψ_p was about 7 bar and growth stopped around 4 bar which represents Y. Since the value of Y could be lowered to 2 bar it seemed to be exerting a regulatory role. Thus m was found to be 0.1–0.2 bar h^{-1} (cf. Eq. [3.4]).

Even more recently Zimmerman and others have performed similar work first on *Valonia* giant algal cells and more recently on *Mesembryanthemum* bladder cells. These cells are unusually accessible cells being exposed on the epidermis and since *Mesembryanthemum* is a flowering plant hopefully results obtained using it would be more generally applicable than work on algal cells.

A new type of pressure measuring device has been used with a microneedle tip. It can be fitted with electrodes so that electrical measurements can also be made. This device (Zimmermann and Steudle, 1974; Fig. 8.4) has already been used to make many

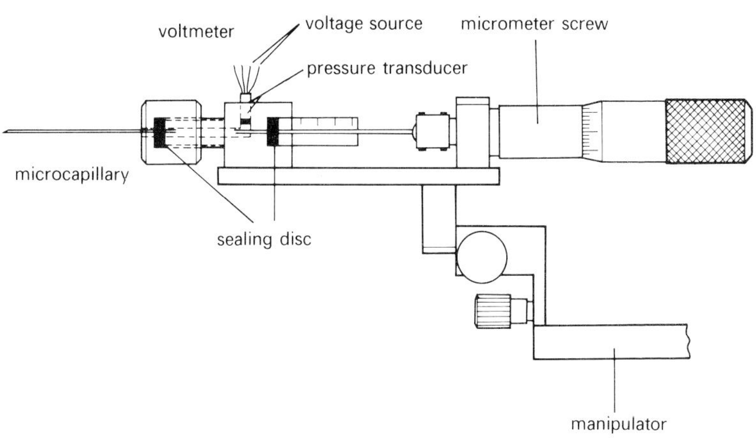

Fig. 8.4 Pressure measurement equipment as used to measure Ψ_p in plant cells. The micrometer exerts a pressure to balance exudation from a plant cell into the microcapillary pushed through the wall using the manipulator. Pressure and voltage can be measured electronically (from Zimmermann and Steudle, 1974).

measurements of pressure and volume changes in cells from which J_v and L_p have been calculated. It has also been possible to measure the cell extensibility.

The latter work illustrates the complexity of growth studies because the hydraulic conductance of cells L_p has been shown to increase as cells approach plasmolysis. Furthermore ion fluxes seem to be influenced by pressure and probably have an important regulatory role during growth. Nevertheless the techniques seem very promising and we may hope to understand the mechanism of growth more precisely in the near future.

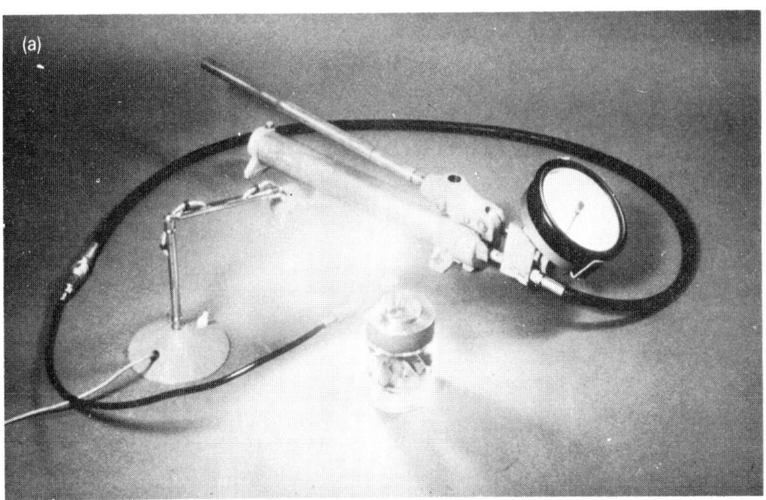

Fig. 8.5 Oil-bomb equipment showing the clear plastic chamber, pressure gauge and hydraulic pump. The chamber is normally illuminated in a water bath (not shown). The petiolar stump projecting from the bomb lid is protected from evaporation by a cap of metal foil.

A new device – the oil bomb

Most of the work on growth described above occurs in response to water made freely available as when discs or leaves are supplied with free water, i.e. at zero water potential. It is obvious from studies using the pressure bomb that the occasions when the water potential is zero within a plant are few and infrequent – the environment in which cells grow is normally a negative water potential and sometimes it can become very negative indeed! The free supply of water to plant cells is not a particularly natural condition; its chief advantage is that water uptake can be measured easily by weight increase or volume changes. On the other hand plants are frequently subjected to restricted water supplies and must change physiologically in response.

If a leaf, still capable of growth, was enclosed in a standard gas-pressurised bomb we might expect to have to increase the pressure to make the xylem sap exude. Either a decrease in turgor pressure through wall relaxation or an increase in solute concentration in the vacuole might produce this effect. Growth would therefore change the xylem sap pressure which would become more negative. In this way the growth capacity of a leaf might be measured, not in terms of weight or volume change at zero water potential, but as changes in water potential with zero water supply.

Two problems raise serious practical difficulties. The first is that a leaf in a gas bomb tends to lose water by transpiration through distillation, caused by small differences in temperature (even respiratory temperature changes would be significant). Clearly if the leaf lost water a change in balancing pressure to produce exudation might be caused, not by growth, but by a normal water deficit. The second problem is that a leaf in darkness produces significant amounts of carbon dioxide by respiration over the lengthy periods we might want to study growth capacity. Both of these problems have been overcome by the invention of an oil bomb. Leaves are immersed in liquid paraffin and kept in a transparent (Perspex) bomb under illumination. In this way water loss is prevented completely while carbon dioxide evolution is reversed by natural photosynthesis. This bomb (see Fig. 8.5) is pressurised with a hydraulic pump and even high pressures do not entail explosion risks as might be the case with gas-filled bombs (Milburn and Dodoo, in prep.).

Using this equipment it has been possible to confirm and extend the picture which had emerged previously from water uptake studies. Thus water uptake by young growing cells of leaves can generate xylem pressures approaching -15 bar even though when taken from the growing plant they have water potentials of around -4 bar. As expected, older leaves change the xylem sap pressures little under the same conditions (Fig. 8.6). We have been able to resolve the question

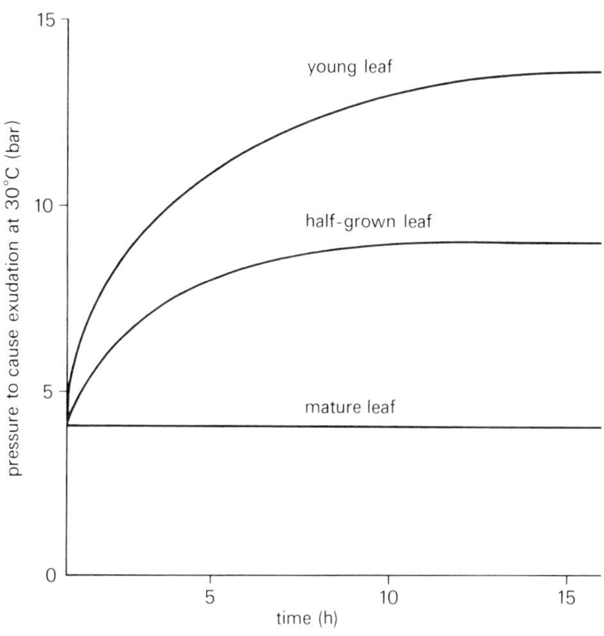

Fig. 8.6 Diagrammatic representation of the course of growth of young, half-grown and mature leaves when enclosed in an oil bomb. Growth is measured as the pressure required to cause exudation (from the projecting petioles) which in young leaves builds up rapidly for several hours but in mature leaves remains constant (from Milburn and Dodoo, unpublished).

Table 8.1 Summary of critical water potentials at which growth of leaf cells is stopped in a range of crop plants. Though the water potential is negative the cell turgor pressures remain positive at the threshold for growth. Ψ is deduced from Ψ_s of leaf cell sap and $\Psi_p + \Psi_s$ of xylem sap determined with a pressure bomb.

Plant	**Water potential** (bar)	**Osmotic potential** Ψ_s (bar)	**Turgor pressure** Ψ_p (bar)	**Author**
Sunflower	−4.0	10.0	6.0	Boyer (1968, 1970)
Soybean	−12.0	−13.0	1.0	Boyer (1970)
Maize	−8.0	−15.0 to −16.0	7.0 to 8.0	Boyer (1970)
Maize	−8.0 to −7.0	–	–	Acevedo, Hsiao and Henderson (1971)
Cotton	8.0	–	–	Jordan (1970)
Ricinus (Castor bean)	−7.5 to 12.0	−8.5 to −17.0	1.0 to 5.0	Dodoo and Milburn (in prep.)

of the mechanism driving growth beyond reasonable doubt. Though large measurable changes in balancing pressure are measured (see Table 8.1), the changes in solute concentration of leaves are insignificant. The only possible conclusion is that cell wall relaxation is the initial stage in growth. Stretching cannot be involved, through pressurisation, because water is not available for this process.

Many new studies have been performed with the oil bomb which is proving a most useful tool for the investigation of plant growth under water stress (Milburn and Dodoo, in prep.).

Growth of organs: a caution

Cells are organised into organs which perform specific functions. Nevertheless the power for growth is provided by individual protoplasts expanding by vacuolation. Controlling expansion is cell wall extensibility but also the extensibility of other cells of the tissue. The normal arrangement in higher land plants is that expansion is provided by an aggregation of thin-walled parenchyma cells forming a pith. Other surrounding tissues brace against the pith in a manner analogous to the way in which guy-ropes of a tent pull against the central tent pole. The main tensioning system in stems and petioles seems to be the phloem fibres. When pith cells have completed their expansion they frequently senesce and become air-filled, the tissue having lost its extensibility.

It is easy to examine tissue tensions by simply cutting organs open. Owing to the natural elasticity in the system the different effects of compression and tensile pull produce curvature. Many different organs have been used, for example the dandelion (*Taraxacum officinale*) flower stalk curves to a Y shape when split longitudinally and for many years physiologists have placed these split stalks in osmotic solutions. When a solution is found in which the split stalk neither opens nor closes it is argued that the water potential equals the water potential of the tissue cells. There are two serious objections to this view: first, the tissue tends to grow and second, the cells after cutting have not necessarily the same water potential as when pressurised in the intact organ. It must be remembered that organs are not merely single cells in aggregate but additional forces come into play which must be recognised. Of course the bioassay based on the split epicotyls of etiolated pea seedlings which curve to different degrees in hormone solution is a quite different application which is entirely legitimate when correctly applied.

The chemical control of growth

No chapter on growth would be complete without mention of the chemical regulation of growth, especially by natural hormones. The literature is vast and detailed inquiries should be directed to modern texts, e.g. Wareing and Phillips (1976).

Certain chemicals have been shown to influence growth very greatly at concentrations so low that they cannot be regarded as nutrients required for the growth process itself. These substances, notably IAA (indolyl acetic acid), GAs (gibberellins, over 50 of which are known), CKs (cytokinins), ABA (abscisic acid) and ethylene are known to occur naturally and each appears to regulate more than one aspect of growth. Thus IAA rapidly affects cell expansion at concentrations below $10^{-5}M$ which reflects the low dosage required for action. Similarly GAs seem to affect cell turgor while CKs seem primarily concerned with cell division and the control of senescence. Additionally, gases produced during growth, particularly ethylene, have pronounced effects on cell turgor and enlargement at extremely low concentrations (below 1 part per million).

The mechanisms underlying the chemical control of growth are difficult to elucidate because the amounts of active chemical are so minute. Furthermore a vast range of chemical inhibitors are known to occur in plant cells and also processes which degrade natural regulators. When, for example, leaf tissues expand in water the rate of expansion seems to be controlled by the production/destruction rates of endogenous hormones. Synthetic chemicals with close chemical affinity can interfere with growth. Thus NAA (α-naphthalene acetic acid), which mimics IAA, produces gross uncontrolled expansion which kills the plant – hence its use as a weedkiller.

Ultimately the complexities of chemical growth regulation will be understood through specific effects on cell wall plasticity, membrane permeability and nuclear activity. This topic will undoubtedly continue to produce controversy and also fascination.

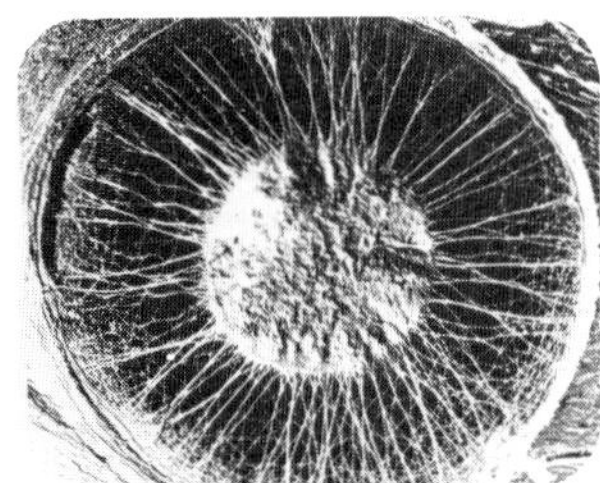

Chapter 9

Plants in adversity

So far we have considered plants under relatively normal conditions, but these conditions may become very adverse, especially in arid or cold regions. To a very considerable extent our capacity to exploit solar energy for the production of food is dependent on our making unfavourable regions more amenable, e.g. by irrigating arid lands. But, it is also necessary to select species and varities of plants able to respond and survive in adversity. The selection of suitable plant types largely devolves on the plant breeder. As yet the main approach has been to test new varieties empirically, selection favouring those with optimal yields under the particular adverse conditions. Such is the phenotypic masking effect of genes, that advantageous features can be obscured by other unfavourable genes and discarded. One goal for physiologists is to identify the limiting physiological processes in survival and to provide the plant breeder with identifiable characters ideal for survival; if selection could be made rapidly, great new advances in plant breeding may be expected. Vast areas of the world are unsuitable for plant production at present on account of aridity, toxicity (including salinity) and frigidity of a semi-permanent nature. Even normally fertile lands may suffer temporarily, e.g. from unprecedented droughts or floods. Also in cultivated crops there are vast losses from pathogenic attack. Many problems depend on modification of the environment utilising known technology; others, especially when questions of feasibility and economic efficiency are considered, depend on an understanding of basic physiological processes in plants.

Plant survival during drought

In recent years considerable efforts have been made to study plant growth in arid conditions, e.g. in the Negev desert in Israel or the Sahara desert. Not all the work has centred on agricultural plants, many attempts have been made to study the characteristics of the naturally occurring plants to provide clues as to the most advantageous stratagems for survival (Stocker, 1974).

Many xerophytes compensate for their arid environment by producing very deep-rooting systems, disproportionately larger than aerial organs compared with typical mesophytes – see Table 9.1.

Table 9.1 Root penetration depths of plants growing in different deserts seem closely related to survival in dry desert environments.

Genus	Location	Root penetration	Source
Alhagi	Asian desert	Depth 25 m	Daubenmire (1959)
Glycyrrthiza	Asian desert	Depth 15 m	Daubenmire (1959)
Andina	Brazilian desert	Depth 19 m	Daubenmire (1959)
Prosopis	Arizonan desert	Depth 20 m	Daubenmire (1959)
Tamarix	Saharan desert	Radius 40 m	Ladover (1928)
Larrea	Saharan desert	Radius 27 m	Ladover (1928)

Doubtless such roots provide a useful store of water, but succulents and cacti excel in this respect. The giant Sahuaro cactus can hold 5,000 kg of stored water in the interior of its specialised stem. Recently attention has been focused by Passioura (1972) not so much on internal water storage but restricted utilisation. He argued that if roots of wheat could be induced to utilise a limited amount of water very sparingly, instead of producing a large vegetative plant, which would die before maturity, a smaller tougher plant could be produced with sufficient water reserves to complete its life cycle, i.e. produce a useful yield. Passioura's interesting experiments in which he grew plants with restricted rooting systems indicated that, although very high Ψ_p gradients must have occurred in the functioning roots, nevertheless the plants did survive and seemed to fulfil his predictions.

The possibility that desert plants can utilise atmospheric humidity or dew deposition under favourable conditions has caused considerable debate. A basic problem is that if a plant is insulated to protect it from desiccation it is difficult to expect transport to reverse at a rate sufficient to provide significant amounts of water. If the water potential Ψ of a plant is, say, -50 bar, the environment, by the combined effects of humidity and thermal gradients could exert a desiccation water potential of say $-3{,}000$ bar (about $\equiv$ 15 per cent r.h.

at 30°C), i.e. a $\Delta\Psi$ of 2,950 bar. Even if this gradient reverses at night, so that it reaches saturation outside the plant, a maximum *reversed* gradient $\Delta\Psi$, of a mere -50 bar, could be established across the critically slow gas phase. Some plants have hairs which enable them to absorb dew directly, but there is no evidence that *useful* amounts of water are absorbed directly from the vapour phase. This is not to say that the process does not occur; indeed a leafy shoot surrounded by tritiated water vapour rapidly becomes radioactive through tritium diffusion via the stomata. What may occur naturally is that water condenses in the upper soil horizons, especially beneath rocks, and this supplies plant roots indirectly, without true rainfall. In effect a flat rock acts as a primitive distillation system because it cools rapidly at night and condenses water from the humid air. Such devices have been used since antiquity; maybe they can be replaced with advantage by more technologically suitable materials such as metal or plastic sheets (see Gindel, 1973).

Water restriction and metabolic processes

It must not be imagined that if water uptake ceases internal water movement in a shoot is stopped. This was illustrated by an experiment on a *Pelargonium* shoot, about 0.25 m long without roots or any liquid water supply, which was suspended in our heated laboratory but away from strong illumination. New leaves were produced from the apex for a few months while the older leaves senesced and abscissed. The stem, normally fleshy, gradually shrivelled suggesting that water was lost from this store, a hypothesis confirmed

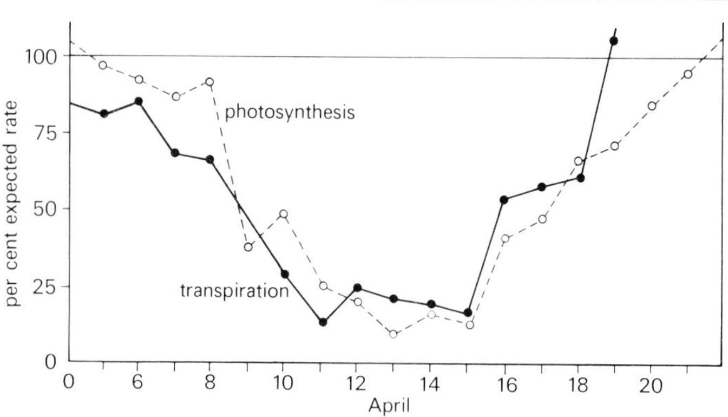

Fig. 9.1 Data illustrating the importance of water availability on photosynthetic assimilation in apple trees. Though it is often difficult to distinguish between stomatal and metabolic effects, it seems that photosynthesis is curtailed by a reduction in cellular water potential (from Heinicke and Childers, 1936).

by reweighing the shoot. After four and a half months the shoot flowered successfully! Subsequently it declined and was assumed dead after about six months. This experiment illustrates the remarkable efficiency of stomatal control in restricting transpiration and the amazing fact that growth can persist in one tissue at the expense of water from other tissues which are sacrificed.

The persistence of growth during water restriction is not particularly surprising. It is well established that plants have the capacity to transport solutes even when severely wilted. *Ricinus* is capable of exuding from its phloem, which must imply that positive turgor pressures are generated in sieve tubes even when the plants are at water potentials less than −20 bar (Milburn, 1975). For extension growth to occur in new organs it is only necessary to raise turgor pressure in a cell to above wall pressure.

A restriction of water reduces growth but also has profound effects on other metabolic processes. For example, transpiration tends to be correlated with photosynthetic assimilation rate (Fig. 9.1). This is partially explained by the stomatal control of gaseous exchange (Fig. 9.2) but partially through internal biochemical changes within the leaf. For further discussions of these complex interactions see Kozlowski (1969).

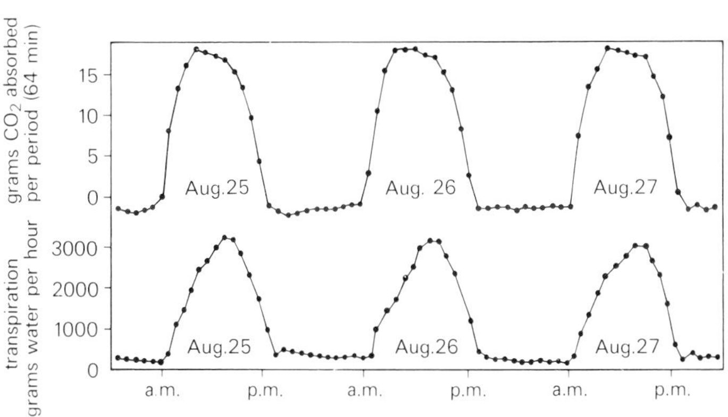

Fig. 9.2 Relationship between transpiration rate and carbon dioxide uptake of an alfalfa crop over a period of three days (from Thomas and Hill, 1937).

Irrigation

Over long periods, plant growth is often a clear reflection of the water supply. This is most obvious when the supply of water is the main factor restricting development. Grasslands in SW. Africa grow in proportion to the annual rainfall (Fig. 9.3). It is remarkable the extent

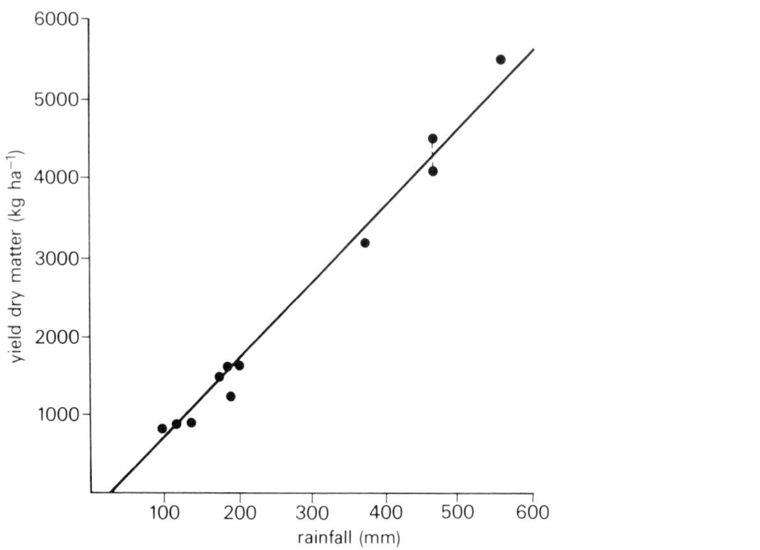

Fig. 9.3 Above ground production of grassland dry matter in kg(ha)$^{-1}$ in SW. Africa is linearly related to the annual rainfall in mm (from Walter, 1973).

to which this growth, expressed in terms of dry matter, i.e. photosynthetic product, is linearly related to the water supply.

An obvious technique practised by mankind since antiquity has been to supplement natural rainfall by artificial irrigation. Thus even cabbages in temperate climates respond to irrigation, but there is a finite limit to enhanced yields obtainable by supplementing water supplies (Fig. 9.4). Note that unlike Fig. 9.3 the results are expressed in terms of fresh (not dry) weight and some of the increased yield (in Fig. 9.4) will undoubtedly represent water content. In agronomy the cost advantage of irrigation frequency must be carefully computed.

Resurrection plants

Plants which can be dried until they seem dead but which revive by rehydration have been called 'resurrection' plants. They are not uncommon in arid regions of the world but represent many unrelated groups, including mosses, pteridophytes and angiosperms, especially grasses. Work on their physiology has shown that many of these plants can survive drying to the point when they equilibrate with an atmosphere near 0 per cent relative humidity, i.e. an almost infinitely low water potential (Gaff and Hallam, 1974). It is not sufficient to show that the plants merely reabsorb water because this might be a purely physical response. Revived plants must be shown to be capable

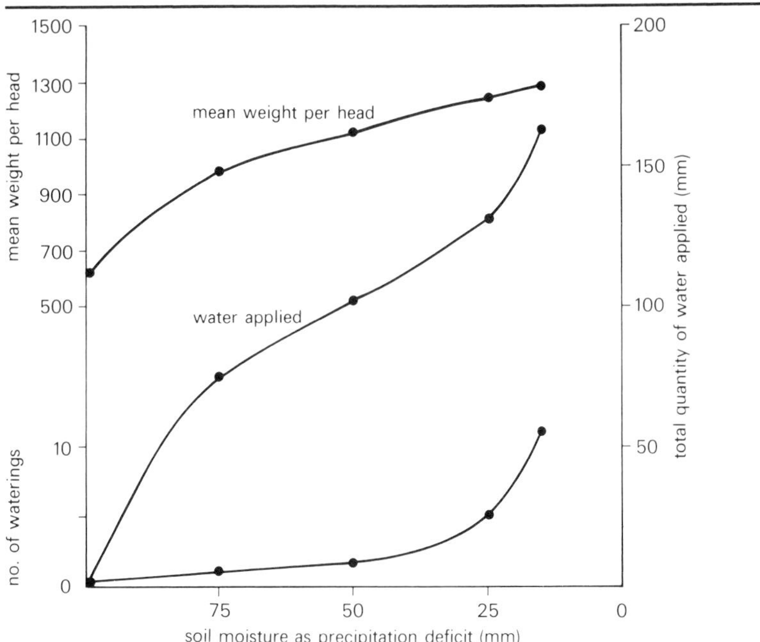

Fig. 9.4 The effect of irrigation (expressed as quantity of water and frequency of application) on mean weight per head of fresh cabbage in grams. There is a law of diminishing returns on benefits from irrigation (redrawn from data of Drew, 1966).

of photosynthesis, for example, Gaff and Hallam have distinguished between resurrection plants which retain their fine structure in shrunken form when dried and others in which membrane structures are lost but reformed when rehydrated. Apparently in these plants the cytoplasm is physiologically very tough. Except in mature plants, this is not so unusual as might be supposed. Seeds of most plants contain highly desiccated embryos which survive with very low water contents. Indeed this behaviour is so commonplace we often forget that drought survival is normal behaviour for most very young plants.

Drought avoidance and survival

Irregular water supplies have threatened plant survival since life on land began and several mechanisms have evolved to overcome the problem. In some plants a secondary reduction in leaf surface occurs through leaf fall or a reduction in size of new leaves. Others, e.g. *succulents* and *cacti*, utilise the leaves or stems as water storage organs and become semi-dormant in adverse periods. Another stratagem is adopted by *ephemeral* plants which germinate following rain and complete their life cycle so rapidly as to survive the subsequent drought as dormant seeds. Plant life on land has always tended to be dominated by plants able to exclude light from their rivals – hence the evolution of massive trees and monocotyledons. The efficiency of a spreading root system to support an immense canopy of foliage exposed on high to atmospheric desiccation is extremely critical.

Several additional devices seem to augment the stomatal mechanism. Sunken stomata found in *Ericas*, the heaths, are probably a modification to regulate wind effects. Leaves sometimes roll to enclose the photosynthetic surfaces through the action of elastic hinge cells which change volume in response to turgor pressure changes. Marram grass (*Ammophila*) is a superb example of this. The outer surfaces of epidermal cells themselves are normally protected by an imperious secretion called a *cuticle* which functions as an insulator. Its role may be augmented further by hairs or waxes which are also outgrowths or secretions from the epidermal cells.

Efficiency of transpiration – anti-transpirants

In arid lands the economics of plant production can be analysed in terms of the efficiency of dry matter production per unit of water consumed by a crop. Estimates are dependent on many variables but representative values are listed in Table 9.2.

There have been several attempts to increase crop efficiency, by decreasing the proportion of water utilised to dry matter produced, using *anti-transpirants*. An anti-transpirant could operate in several different ways. There is little likelihood that overheating caused by a reduction in transpiration would be damaging; desiccation seems the more serious problem to plants. Ideally a film of plastic material applied by spray might veil the stomatal apparatus making it permeable to carbon dioxide but relatively impermeable to water

Table 9.2 The efficiency of water used W/D expressed as the weight of water W required per weight dry matter produced D for different types of crop.

Plant type	Growing conditions and physiological system	Water efficiency W/D
Mesophytes	High yielding crops	2,000 and over
Mesophytes	Water restricted (C_3 plants)	*c.* 700 ± 250
Mesophytes	Water restricted (C_4 plants)	*c.* 300 ± 50
Succulents	CAM plants (PEP carboxylase)	0.55 and below

vapour. Alternatively an oil might be applied which would form a surface film providing a barrier to the diffusion of water vapour. Considerable savings of water have been made in Australia by coating the water surface of open reservoirs with a surface active material. As yet such physical techniques have proved very expensive to apply to plants and are largely unsuccessful.

A more promising method is to attempt chemical control by inducing stomata to close partially (see Ch. 7) and so conserve water throughout the growing season or in the face of impending drought. Inhibitors which do induce stomatal closure have tended to reduce crop yield in amounts which negate the expense of their application. Nevertheless there is hope that control by natural hormones or their chemical analogues may one day provide a solution to this important problem. Another possibility is that an anti-transpirant might induce premature leaf fall. The deciduous habit is an effective method perhaps to reduce transpiration in adverse seasons (especially freezing), but whether it can be controlled effectively by man is open to question. Perhaps a more mundane but remarkably effective method for conserving water in arid lands is the use of glasshouses. In temperate climates glasshouses are used to retain heat, but in the tropics their principal role would be to minimise the loss of water vapour in proportion to the uptake of carbon dioxide. Pilot plants are in operation in the Persian Gulf and phenomenally good productivity figures have been quoted.

Succulent physiology

Scientific detection has revealed another fascinating plant stratagem called crassulacean acid metabolism (CAM). Over a century ago it was discovered that the fleshy tissues of genera of the Crassulaceae (e.g. *Bryophyllum*, *Sedum*) became markedly more acid by night, up to 1,260 per cent, measured by titration. Later it became clear that many succulents (e.g. *Opuntia* and pineapple) behave in the same way and two additional clues emerged. They were found to open their stomata by night and close them by day (the converse of most plants) and their tissues had a remarkable capacity to absorb carbon dioxide, coinciding with acidification. Apparently CAM plants are ideally adapted to store water and their physiology is adapted to dry environments. Carbon dioxide is accumulated by night when it is lost from non-CAM plants by respiration. The following day the carbon fixed as organic acids is converted photosynthetically while the stomata are closed to restrict water loss. The advantage of this system may be judged from the growth of pineapple, a CAM plant, over sugarcane (a C_4 plant) which is not. To produce the same increment of dry matter sugarcane uses about ten times more water (Ekern, 1965) and C_3 plants would require even more water.

Very recently the fascination of CAM plants has taken a new twist. It has been reported by Hartsock and Nobel (1976) that when water-deficient CAM plants (*Agave deserti* and others) are watered daily, after a few weeks their behaviour pattern shifts. In the new pattern the stomata open in light, the normal daytime pattern of most plants, and the maximum rate of carbon dioxide absorption also shifts from night to day. This physiological device seems to increase the efficiency of carbon dioxide absorption by CAM plants to that of other C_3 and C_4 plants, making them more competitive under favourable conditions.

Mangroves – desalination plants

Studies on the osmotic potentials Ψ_s of mangrove saps by Walter and Steiner (1936) gave an indication that mangroves might be subject under normal conditions to quite severe internal water stress. Scholander *et al.* (1968) have extended this work. Mangroves normally grow in estuaries or marine situations with their roots periodically or continuously bathed in seawater (Ψ_s approx. -25 bar) from which they must extract any water lost in transpiration. Nevertheless the xylem sap, expressed by pressurising shoots in a pressure bomb, proved to be virtually fresh water ($\Psi_s \approx 0$). In contrast the osmotic pressures of the cell sap from cells in mangrove leaves proved to be very considerable which was balanced by considerable xylem sap tensions Ψ_p

Type of plant	Genera (examples)	Range Ψ_p xylem sap
Mangroves	*Rhizophora, Osbornia,*	-25 to -50 bar
Other Halophytes	*Salicornica, Suaeda,*	-40 to -65 bar

When the uptake of water was prevented artificially mangrove shoots developed xylem tensions exceeding -70 bar, showing very clearly that in principle the mangroves could desalinate seawater by ultrafiltration. Experiments in which mangrove roots immersed in seawater were enclosed in a pressure bomb showed clearly that under external positive pressures, corresponding with normal xylem sap tensions, the roots exuded almost fresh water. A similar process, technically 'reverse-osmosis', utilising artificial membranes and positive pressure systems is used for purifying seawater.

Mangrove roots were found to possess an ultrafiltration system with a reflection coefficient σ of 0.95 for salts, which behaved like a physical membrane being remarkably insensitive to inhibitor poisons such as carbon monoxide and dinitrophenol. Despite the fact that some salts penetrated the roots, sap expressed from shoots closely approached pure water, suggesting that a secondary mechanism of salt removal occurs through salt excreting glands which are present on leaves of many mangroves. By these remarkable adaptations, mangroves are able to flourish in habitats so adverse as to exceed the tolerance of most plants.

The effect of freezing temperatures

Plants are liable to damage when the water within them undergoes freezing. At the subcellular level if protein solutions are frozen they may be denatured, and this presumably occurs during frost damage. Possibly more serious is the growth of ice crystals which cause serious disruption by effectively harpooning the cell membranes. Damage proceeds when thawing takes place because the biochemical control of metabolism has been disorganised. Both processes can be reduced under laboratory conditions by very rapid freezing which reduces crystal size to less disruptive dimensions.

If many cells (e.g. those of *Brassica* spp, cabbages and Brussels sprouts) are allowed time to adjust to cooler conditions they become 'hardened'. One of the main protective devices is the capacity to allow intracellular water to escape and freeze in the extracellular spaces where the crystals are less disruptive.

Inside cells the contents are osmotically active and the partially structured water resists freezing. In the laboratory cells can be treated with cryoprotective solutions such as glycerol which protect them in a manner somewhat analogous to plasmolysis effects. The familiar drooping of *Rhododendron* leaves in cold conditions is caused by the escape of the aqueous vacuolar contents from cells in the petioles. If such petioles are warmed the cells reabsorb water and recover, even without an external water supply.

When gas-saturated water freezes the gases are expelled as bubbles which may often become trapped in the ice. What effect would bubbles have on a conducting system operating under negative pressures such as the xylem? It seems that the effects are disruptive, as might be supposed. Damage is restricted in a number of ways. In a finely porous wood (with many fine tracheids) as freezing takes place seems that the first gas bubbles may 'seed' gas from the surrounding sap so that when freezing finally occurs only a small proportion of conduits are gas filled and the system is able to conduct, though in an impaired condition. This seems to be the system operative in many gymnosperms, e.g. *Pinus* (pines), *Abies* spp (firs), and diffuse porous hardwoods, e.g. *Acer* (maples) and *Populus* spp (poplars).

In trees with wider, longer conduits the capacity to survive freezing damage is probably less, the conduits are vulnerable on account of their size because a single bubble might disrupt the system. The main mechanism to circumvent frost damage in these ring-porous trees is to produce a completely new set of sap-filled conduits from the cambium immediately before the growing season. It is for this reason that in trees like *Quercus* (oak) and *Fraxinus* (ash) practically all xylem conduction in summer occurs via the outer current growth ring of xylem, but this is to some extent dependent on the severity of the preceding winter.

Some trees like *Betula* (birches) and creepers like *Vitis* (grapevine) seem to have yet another protection system available to repair frost embolised conduits, because they develop considerable root pressure immediately before bud break in the spring. While it is most likely that a column of pressurised sap could expel air from virtually dry xylem conduits, little work seems to have been performed to test this possibility.

The phenomenon of leaf-fall, so obvious in temperate regions, is an adaptation to meet the onset of winter. Evergreens, e.g. pines, also exhibit leaf-fall towards the end of the growing season, but on account of the foliage which persists this can give rise to the false impression that they retain their leaves. From the fact that evergreens and deciduous plants compete in temperate regions, it seems that the advantage of rapid growth of vulnerable systems may be balanced against slower growth of more resistant but less efficient tissues.

Water stress and cavitation

Opinion on the mechanism of sap ascent in plants has fluctuated throughout the last century. Originally it seemed impossible for tall trees to overcome the pull of gravitation on sap because a gas phase would develop (cavitation). In nature vibration, produced by wind, would make water under tension even less stable. Later, after Strasburger, Dixon and others had shown convincingly that water could withstand surprisingly severe tension without cavitation; it seemed that trees might cavitate infrequently, though it was established that as wood aged over the years it gradually became filled with air. Then Gibbs (1958), from numerous measurements on the air content of timber of a range of species at different times of year, suggested that not only did the conduits seem to empty of sap in summer but they also seemed to 'refill' in autumn and winter months.

Surprisingly, quite severe sap tensions develop in small seedlings, crop plants and herbs. Soil moisture and the rate of transpiration are more crucial than gravity on these plants. Milburn and McLaughlin (1974) found that cavitation, detected acoustically, occurs surprisingly readily at tensions of 5–15 bar in herbaceous species, such as *Plantago major* and *Tussilago farfara*, under field conditions. Similarly in *Ricinus*, the castor bean, sap tensions are commonly around 5 bar when the plant is well supplied with water and illuminated. Only a slight restriction in water uptake induces cavitation which can be heard by amplification as 'clicks' even before the leaves wilt. After wilting, if a plant is rewatered it seems to recover completely. This is an illusion however; the plant has suffered severe internal embolisms following cavitation, which may be revealed if the leaves are subjected to a restricted water regime a second time. Cavitation is now practically eliminated until tensions exceed those developed previously (Milburn, 1975). These results are shown schematically in Fig. 9.5. Very similar experiments by West and Gaff (1976) have shown that cavitation occurs at relatively low sap tension (12 bar upwards) in apple leaves. These results support Gibbs's notion that cavitation occurs rather readily in trees. This being so, we must re-examine the question of the reversal of cavitation. This can be accounted for quite easily in small herbs which are capable of generating positive root pressure, which would, combined with the effects of the surface tension of water, rapidly redissolve gas emboli. However, root pressure and other internal mechanisms are quite inadequate to

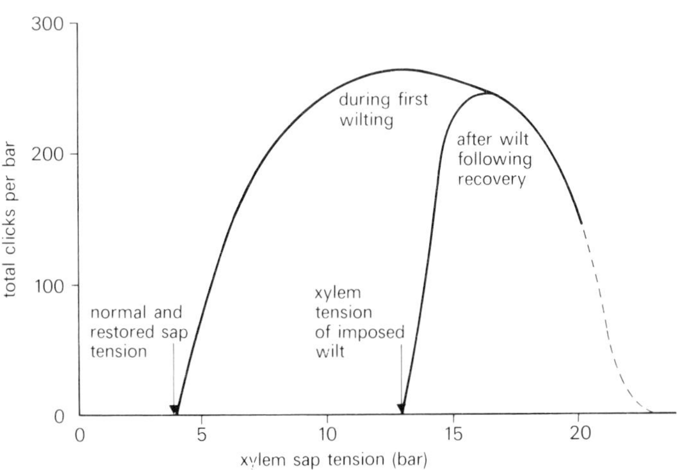

Fig. 9.5 Schematic relationship between cavitation (expressed as 'clicks' per bar) and Ψ_p in *Ricinus* leaves from a plant following restriction in water supply until wilting had been induced. After the plant had recovered and seemed normal, cavitation in a second leaf was significantly less than originally, because only the more resistant conduits were sap-filled (from Milburn, 1975 and unpublished).

reverse cavitation in tall trees. The fascinating problem is how it might occur.

Quite recently the emphasis has again switched to the availability of external water. Plant surfaces are more or less cuticularised, rendering them waterproof. Some uncuticularised areas might absorb rainwater directly. More intriguingly, it seems possible that wind, far from inducing cavitation, may actually produce damage to the cuticular surfaces by abrasion and allow the direct absorption of rainwater, even in the uppermost branches of tall trees. Sap tensions have often been found to fall dramatically during rainfall but this has been attributed to rainwater contamination or root uptake. Current experiments indicate that sap tensions fall sufficiently to allow gas emboli to redissolve through surface tension and so reverse cavitation. The direct absorption of water occurs despite some air embolism at the wounded surfaces by essentially the same mechanism as flower stems which, though cut in air, also absorb water (see Ch. 5). If the absorption of superficial water is established as an important and widespread mechanism many previous unexplained morphological features of plants, such as their surfaces, branching patterns and conduit dimensions and frequency, may be revealed as beneficial adaptations which counteract damage from water stress.

A standard approach to gauge the water requirement in a given season is to measure the difference between the quantity of water evaporating from a reference tank of water and that collected from rainwater. Empirical formulae have been developed to take into account such factors as unusually heavy rain which may exceed the absorbance capacity of a soil and so be lost as 'run-off'. There is an increasing use of electronic instruments designed to model a plant, so constructed that irrigation is automatically switched on when the model plant begins to dry. Another recent detector utilises infrared photographs taken from artificial satellites. Dry zones tend to radiate more heat than those in which plants are cooled by transpiration, and this can be detected on a truly global survey. As water supplies become more restricted and food production more important, the cost benefits of water must become a crucial issue in the efficient harvesting of world resources.

What is adversity?

The perception of adversity to plants depends on our understanding. Gradually we have unravelled many potential hazards to plant survival such as unrestricted water loss and the mechanisms (stomata. cuticular secretions, etc.) which have been evolved to control it. We now understand why CAM plants tend to close their stomata in sunlight. The wonderful complexity of vascular systems can be appreciated only when the dangers of embolisms in xylen or excessive phloem bleeding are appreciated. Undoubtedly many unusual systems await explanation.

No satisfactory explanation has been given for the remarkable behaviour of sensitive plants of the genus *Mimosa*, the leaves of which droop when touched, but recover turgor in less than an hour. The leaves also adopt this 'sleep' position in darkness which may provide a clue to the reason for this behaviour. Many other plants such as wood sorrel (*Oxalis acetosella*) or the seedlings of maple (genus *Acer*) alter the orientation of their leaves in response to high or low levels of illumination. If carefully observed, a *Mimosa* plant can change leaf positions continually in response to sunlight throughout the day, just as the sails of a schooner might be adjusted to suit the prevailing wind. Is the sensitive response merely a by-product of a highly evolved mechanism which stabilises photosynthesis or is it a device to control leaf water relations? We do not yet know.

The assumption that any complex device or behavioural pattern serves a beneficial function to the plant seems to me fundamentally reasonable. It is wrong to claim that a plant possesses a structure *because* it is beneficial because this, by teleological argument, implies the plant consciously seeks some desirable objective. Nevertheless if the products of evolution survive a prolonged selective process which continually rejects unsatisfactory or wasteful designs and useless deadweight any elaborate structure is difficult to explain. Consequently while teleological arguments cannot be defended rationally they may indicate designs which have been evolved to counteract adverse conditions.

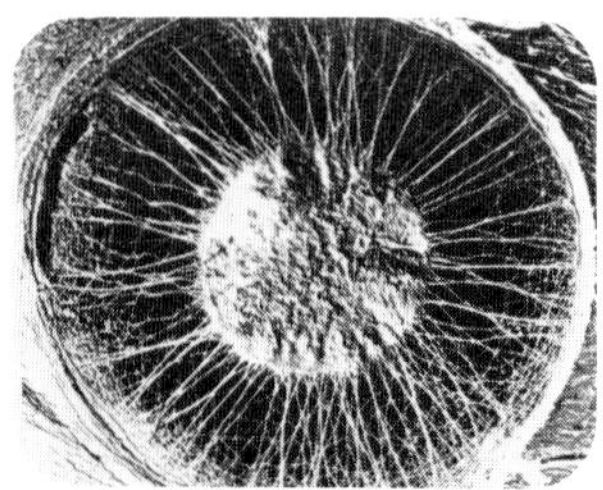

Chapter 10

Ancillary systems and exudation phenomena

Introduction

Until now we have concentrated on flow of water through the xylem, which must be regarded as the main irrigation system within a plant. There are, however, other channels of transport, and it is important that their function be considered in relation to water transport. We know from observations on exudation that other channels are often capable of conducting water in bulk, by pressure flow. The extent to which they function as significant transport systems in whole plants is not easy to decide. We will consider in this chapter phloem, latex ducts, resin canals and ancillary xylem exudation including maple sap flow and pathological xylem exudation. Root stump exudation, due to osmotic uptake, has already been dealt with in Chapter 4.

Bark exudation

Phloem transport of water

According to Münch's pressure-flow hypothesis, flow through sieve tubes can be considered to be the resultant of two osmometers A and B connected so that (Fig. 10.1) their internal pressure is common. The internal pressure at equilibrium corresponds with the extent to which solutes in the osmometers lower the water potential below that bathing the osmometers. When equilibrium is disturbed, for example by a release of solutes in osmometer A, it becomes a source, and an

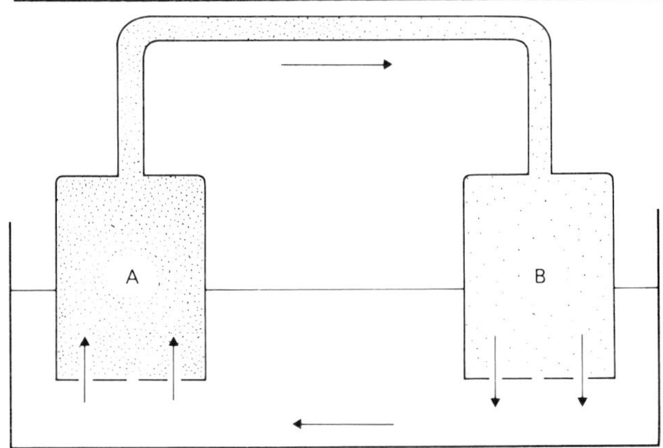

Fig. 10.1 Münch's model to illustrate phloem transport. Two osmometers with different Ψ_s but common Ψ_p act in opposition. Consequently water enters one osmometer but is expelled from the other, producing a bulk flow of water which carries solutes passively (mass flow).

internal gradient of water potential is set up so that there is no longer balance across the membranes of A between water potentials due to solutes and the internal pressure. Consequently there is an influx of water at A causing the pressure to increase, which in turn causes expulsion of water from the sink B by reverse-osmosis. The flow of solution induced from A to B is called mass, bulk or pressure flow. Solutes tend to accumulate at B until equilibrium is restored, but if we imagine that neighbouring tissues absorb such imported solutes the process of bulk transport of solutes and water could continue indefinitely.

What interests us in this section is the expulsion of water. Münch proposed that, since the xylem sap is normally under tension, any water released would be recycled in the xylem. He predicted that as much as one-tenth of the total xylem water could be recycled in this way. He attempted to prove his theory by cutting bark flaps from a tree but leaving them still attached from above, and indeed collected quantities of water in waterproof envelopes from his artificial 'sinks'. It is, however, a common observation that most plant organs enclosed in a plastic bag release water by distillation which can be collected, and Münch's (1930) experiments in no way proved that water collected from bark flaps exuded as *liquid water* in accordance with his theory.

Recently Münch's experiments were repeated on bark flaps, and these proved quite conclusively that liquid water was indeed produced by exudation (Fig. 10.2). In this case the bark was enclosed in tubes

Fig. 10.2 Tongue of *Salix* bark still attached to the tree but immersed in a tube of paraffin oil to prevent water distillation. Water droplets exuded from the surface of the bark (see also Milburn, 1975).

filled with inert mineral oil to eliminate the gas phase required for distillation. Under these conditions liquid water collected in much smaller amounts than might have been predicted from Münch's experiments, but the results undoubtedly support Münch's hypothesis.

It is useful to reconsider the amounts of water we may expect to be exuded from the phloem. Münch's idea that one-tenth of the xylem flow could be recycled water is based on the notion that all sieve tubes conduct sap in unison and in the same direction. In fact evidence has grown that sieve tubes often conduct sap in opposite directions simultaneously (e.g. Turgeon and Webb, 1973). If, for example, half the sieve tubes imported sap, and the other half exported the same sap, deprived of those solutes required by a growing organ, water circulation would be confined exclusively to the sieve tubes.

Structure of the sugar pipeline

Sieve tubes are sieve elements connected end-to-end resembling the xylem tracheidary conducting system. Just as pit membranes between tracheids protect the system from gas embolisms, sieve plates, structures between sieve elements, also serve a defensive function. They operate quite differently, however, because pit membranes protect a negative-pressure system through the physical operation of aqueous surface tension. In contrast sieve plates protect a positive-pressure system by preventing the escape of sufficient of the nutritionally-rich sap to sustain dangerous predators. The sealing action depends in part on a massive surge of proteinaceous materials which physically plug pores in the sieve plates. Several additional devices support this mechanism by metabolic deposition of callose or protein which also block the sieve-plate pores; some plants have proteins which coagulate on exposure to atmospheric oxidation. Probably all these sealing mechanisms are reversible under appropriate conditions.

The low osmotic potential Ψ_s of phloem sap, which forms the basis of Münch's hypothesis, suggests that in sieve tubes turgor pressure Ψ_p must be considerable. Yet most sieve tubes when wounded do not exude detectable quantities of sap, thanks to the remarkable efficiency of the sealing mechanisms. Nevertheless sieve tubes of some plants have been known to exude when cut, especially palms and large deciduous trees. Many more convenient experimental plants have been added to the range available in recent decades including *Cucurbita*, *Yucca*, *Tropaeolum*, *Ricinus* and *Lupinus*. Experiments on these plants show that the capacity of sieve tubes to seal varies between species, varieties and even between different parts of the same plant. These observations, in conjunction with other experiments, for example using radiotracers, have largely explained the puzzle set by microscopists. Since the nineteenth century it has seemed that sieve-plate pores were blocked when examined and this view was extended by electron microscopists. Though we do not fully understand the role of the proteinaceous filaments observed in sieve tubes it is clear that sieve-plate pores plug rapidly during fixation for sectioning. Better techniques show much less blockage. Sieve tubes near an exuding cut *must* have perforated sieve-plate pores to permit the observed flow (see Fig. 10.3). It is very difficult to believe that this capacity of sieve tubes to conduct sap is in any way unusual in the intact plant.

The operation of sealing mechanisms in nature is a fascinating topic. Aphids are 'sneak-feeders', withdrawing sap gently so that

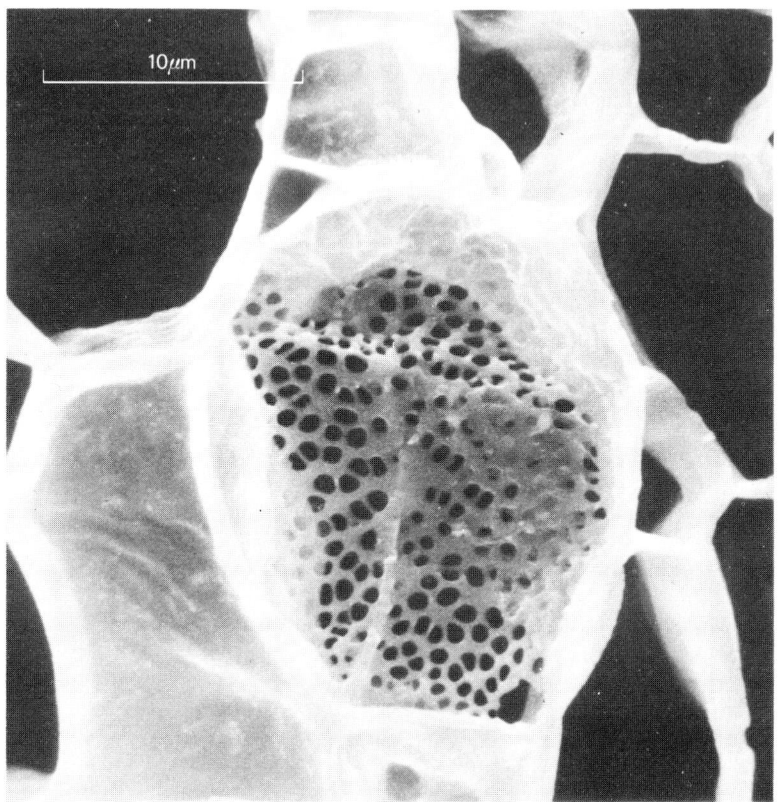

Fig. 10.3 Sieve plate showing open sieve-plate pores from a cross-section of a *Ricinus* petiole. It was cut with a razor blade by hand – the same technique used in exudation studies. The section was freeze-dried then prepared for scanning electron microscopy (from Milburn and Sprent, unpublished).

sealing is not triggered. Aphid stylets continue to exude after the body of the insect has been removed showing the passive role of the insect and demonstrating sieve tube Ψ_p directly. Some phloem predators like the sap sucker, a species of woodpecker, tap trees very systematically by pecking the bark each day. They occasionally kill their host trees. Recently even large animals have been found to be phloem feeders, for example a species of Amazonian pygmy marmoset (Coimbra-Filho and Mittermeier, 1976) cuts into sieve tubes with its teeth which are specially adapted as tusks. Man is himself the largest 'predator', the palm sugar industry alone producing *c.* 100,000 tonnes of sugar annually.

Osmotic effects on sieve tubes

Münch's hypothesis would predict that changes in the phloem water supply should modify Ψ_p. Such changes in pressure should also modify the rate of phloem sap exudation and just such changes were demonstrated by Weatherley *et al.* (1959) using aphid excised aphid stylets. If the xylem water were replaced with osmotic solutions to change Ψ_s the rate of sap exudation decreased. An alternative technique is to reduce (then restore) the water supply to an exuding plant so altering Ψ_p. This was shown very convincingly in experiments on exuding *Ricinus* shoots as shown in Figure 10.4. If the shoot were removed from its xylem water supply the rate of exudation fell sharply but was restored each time water was resupplied. Curiously, very similar demonstrations were made long ago in the thirteenth century by Marco Polo in Ceylon. He observed that phloem sap exudation from a palm tree was restored by watering the roots.

It is necessary to interpret osmotic effects on phloem with caution. Sieve tubes can develop high concentrations of solutes in a matter of hours, so maintaining Ψ_p even in wilted plants. We know this because wilted plants retain the ability to exude, and indeed water conduction in the phloem may survive water transport in xylem when it is disrupted by cavitation.

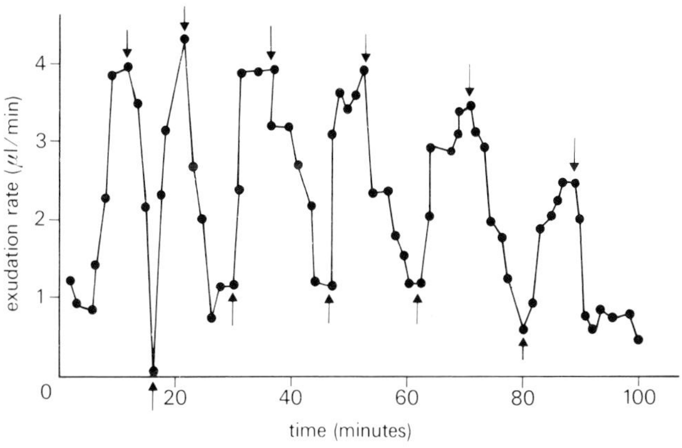

Fig. 10.4 Exudation from razor blade incisions in *Ricinus* bark of a detached shoot. The rate of exudation increased dramatically each time the water supply was restored to the xylem at the shoot stump but decreased when the water supply was withdrawn. The shoot stump was trimmed on returning it from air to water to remove xylem, filled with air emboli (adapted from Hall and Milburn, 1973).

Pressure gradients in the sugar pipeline

In recent years there has been increasing success in measuring the turgor pressure of sieve tubes. Most of these measurements have been made using bubble manometers to measure the pressure. A bubble manometer is attached to a fine hypodermic needle, with a tip specially modified so that it does not easily become blocked when it is pushed into bark. Only a small amount of sap need enter the bubble tube to produce compression (Fig. 8.3). The bubble is compressed in length which is calibrated using pressure equipment.

By applying this technique to different parts of an oak tree Hammel (1968) measured Ψ_p gradients driving phloem sap transport through oak tree bark around 0.44 bar m^{-1}. It is easy to measure pressures less than the true value because there may be leakage at the needle tip or blockage within the sieve tubes before compression is complete. Nevertheless gradients around 0.5 bar m^{-1} are not unreasonable in trees, as indicated by Ψ_s measurements which are described below.

More recently similar experiments have been performed by Sheikholeslam and Currier (1977) on squirting cucumber (*Ecballium*) plants. These plants were about 40 cm tall and pressures were measured near the apex, at the mid-point and at the base. Extreme pressure differences were measured when leaves at the mid-portion were 'fed' with carbon dioxide and light and either the apex was darkened – about 3.8 bar, or the base was darkened – about 5.1 bar. Since these pressure differences were exerted over about 0.2 m the pressure gradients indicated are amazingly high, 19.0 and 25.5 bar m^{-1} respectively!

Against these measurements we can compare Ψ_s measurements on sap collected from wounds in phloem in different positions. These indicate Ψ_s gradients of 0.1 bar m^{-1} or greater in trees but greater gradients from 5 to 10 bar^{-1} have been measured in *Ricinus* plants 0.5 to 1 m tall (Milburn, 1974). To interpret this data in terms of pressure we must assume that $\Psi_p \approx \Psi_s$ as the sap moves along the pressure gradient, absorbing water as it moves along the tube. When the flow is rapid it is unlikely that equilibrium is reached and this method would underestimate the true pressure gradients.

Measurements of sap flow

Two methods are commonly used to measure sap flow. The first method depends on wounding but has the advantage of being direct and simple; the second can be applied to whole intact organs but depends more on assumptions.

Direct method Some plants exude phloem sap sufficiently freely when severed with a sharp blade to permit sap flow to be measured directly in calibrated capillary tubes or even measuring cylinders. These plants include several palms, *Fraxinus*, *Yucca*, *Ricinus* and *Lupinus*. Some of the most spectacular measurements of this kind have been made on sap flow from *Arenga* palms by Tammes (1952). The palm inflorescence stalk was specially prepared by removing the inflorescence. A thin slice was then cut from the stump (still attached to the palm) each day. The flow of sap, consisting of about 14 per cent sucrose solution, built up to 0.24 litres per hour through a measured cross-sectional area of phloem of $3.4 \times 10^{-5}\,\mathrm{m}^2$ ($0.34\,\mathrm{cm}^2$)! Since $J_v = V/At$ (see Eq. [1.6]) the hydraulic flux (which equals the mean sap speed) through the phloem was $1.96 \times 10^{-3}\,\mathrm{m\,s^{-1}}$ (or $70.6\,\mathrm{m\,h^{-1}}$)! Assuming that the pressure gradient driving flow was $1\,\mathrm{bar\,m^{-1}}$ the hydraulic conductivity L was $1.96\,\mathrm{m^2\,bar^{-1}\,s^{-1}}$ of whole phloem, only some of which consisted of sieve tubes. (Estimates commonly vary from 20 to 50 per cent of sieve tubes per phloem cross-section.) It must be admitted that these measurements are probably not relevant to intact, unwounded plants, nevertheless they supply remarkable evidence of the capacity of the sugar pipeline to conduct sap!

Mass transfer sap measurements Another means to measure sap flow through sieve tubes is by measuring the accumulation of mass in the sink organ. This measurement called mass transfer (MT) can be made easily by estimating the dry-weight increase of storage organs, fruits, etc. To make these measurements comparative, mass transfer is expressed per transverse sectional area of phloem (or ideally of sieve tubes). This measurement is called the specific mass transfer (SMT). If the time taken for an organ to store dry matter, the phloem cross-section, and the sap composition (weight per unit volume) are known the hydraulic flux will be:

$$J_v\left(=\frac{V}{At}\right)=\frac{\mathrm{SMT}}{c} \qquad [10.1]$$

Hydraulic flux	J_v	$L^3L^{-2}T^{-1}$	$\mathrm{m\,s^{-1}}$
Specific mass transfer	SMT	$ML^{-2}T^{-1}$	$\mathrm{kg\,m^{-2}\,s^{-1}}$
Sap concentration	c	ML^{-3}	$\mathrm{kg\,m^{-3}}$

On this basis Dixon and Ball (1922) measured the growth of a potato (*Solanum tuberosum*) which increased in fresh weight by 0.21 kg in about 100 days. The dry matter (mainly starch) of the potato was about 24 per cent of the final fresh weight and the phloem cross-section in the potato stalk (stolon) was estimated to be $4.22 \times 10^{-7}\,\mathrm{m}^2$ ($4.22\,\mathrm{mm}^2$). Assuming that the sap contained about 10 per cent sucrose (weight per volume) the volume of sap moving into the potato to carry the measured dry matter (0.050 kg) was $0.05 \times 10^{-3}\,m^3$ (i.e. 0.5 litre) in 100 days which gives a J_v value of $1.37 \times 10^{-2}\,m\,s^{-1}$ ($=0.494\,m\,h^{-1}$) and the SMT equals $1.3 \times 10^{-9}\,kg\,m^{-2}\,s^{-1}$ ($=4.94\,g\,cm^{-2}\,h^{-1}$). Subsequent experiments have supported the validity of this simple experiment. Of course it must be remembered that this calculation gives the average speed of sap flow – the real flow may in fact fluctuate. Furthermore it assumes that all the sieve tubes conduct in the same direction, which may not be true all the time, and also that the loss of carbon dioxide through respiration is zero which cannot be so.

One further fascinating conclusion may be drawn from this experiment. The potato required 0.5 kg of sap to carry the dry matter but contained only 0.16 kg of water when harvested. What happened to the 0.34 litre of water? It is presumed, but not proved, that it was recycled back along the xylem to supply the potato plant.

The pressure gradient driving flow in *Solanum* stolons has not yet been measured. But if it is assumed to be about 10 bar m^{-1} (to approach *Ricinus* and *Ecballium*) then the hydraulic conductivity L of whole phloem can be stimated (from Eq. [10.1]) to be 1.37×10^{-3} $m^2\,bar^{-1}\,s^{-1}$ (cf. $1.96 \times 10^{-3}\,m^2\,bar^{-1}\,s^{-1}$ for *Arenga*). These figures are only applicable to whole phloem and ideally should be refined to apply to sieve tubes alone by measuring the fraction of sieve tubes in the total cross-section of phloem.

In this brief outline of phloem function, the role of water, which forms 80 to 90 per cent of phloem saps and flows in mass or bulk flow through sieve tubes, has been emphasised. For other aspects of this extensive subject see Zimmermann and Milburn (1975).

Latex canals

While sieve tubes and xylem conduits are obviously necessary functional components of a plant, latex conducting tubes do not have any obvious function. Latex is usually milky in appearance and complex in chemical composition. It is collected in considerable amounts from *Hevea brasiliensis* forming the important natural rubber industry. But it must be appreciated that not all plant latex is contained in tubular systems. In many plants latex is contained in isolated secretory cells called *laticifers*, which occur in many different tissues. Only in trees such as *Hevea* are these units anastomosed to form a connected system of *articulated laticifers*.

Rubber trees are tapped by slicing diagonal grooves into these laticifers, which are most abundant in the secondary phloem of the bark. The sap is under considerable pressure and bleeds for a time, eventually ceasing in a manner somewhat analogous to phloem exudation. What is clear is that the tree actually reacts when tapped, producing more latex in response to tapping. This response has been enhanced very significantly in recent years by treating the trees with synthetic hormones, particularly 2,4D naphthalene acetic acid, 2,4,5T and also Ethrel (which releases ethylene), greatly enhancing the yields obtainable and hence the productivity of the industry.

Latex pressures have been measured by inserting small sealed glass tubes containing a bubble of gas (see Fig. 8.3) which is compressed by the latex as it exudes into the tube (Buttery and Boatman, 1966). Pressures taken in the early morning were found to range from 7.9–15 bar, falling by day and increasing by night, evidently influenced by the overall water balance of the whole tree.

Latex contains a mixture of constituents and varies from plant to plant. In *Hevea* it contains isoprenoid hydrocarbons in particular, which burst from vesicles when the latex canals are punctured. Other substances such as terpenes, proteins, alkaloids, waxes, enzymes and starch are found also. There is no evidence that latex canals perform useful translocation in the intact plant or that the constituents are available for reutilisation from storage. Nevertheless the fact that latex is widespread in many plants and that its production can be enhanced suggests it may be of survival value in many plants. A clue to its possible function is provided by Schery (1957) who noted that some trees (e.g. *Castilla elastica* or *Hancornia speciosa*) can be drained of latex completely by a single massive collection. Afterwards they tend to fall prey to boring insects. Possibly, latex ducts, either in isolation or combined into canals, form a defence system against sucking and boring insects through their capacity to seal small wounds.

Resin, gum and mucilage canals

Resin ducts occur in many plants, especially in coniferous trees, where they are present in several different tissues. The ducts form a canal system produced when a number of cells are dissolved to leave a tubular system of cavities, often lined by secretory cells. Thus while exudate from sieve tubes and laticifers originate from the cell interior (being roughly equivalent to the vacuole of a normal cell), resin escapes on wounding as a thin fluid which is strictly extracullular in origin. Resin canals are most abundant in bark but occur widely throughout the plant. On exposure resin forms a hard deposit which effectively plugs and protects wounds.

In recent years it has been found that resin collections, tapped in a manner similar to that used for latex for the production of rubber, can be enhanced by as much as 46 per cent by treating the wounded bark with *c.* 40 per cent sulphuric acid or 2,4 D naphthalene acetic acid (Clements, 1967). Apparently these chemicals act by enhancing the wound stimulus (see Velkov and Kaludin, 1970).

Resin canals are clearly capable of water conduction by pressure flow when the system is wounded, but there is no indication that they serve for conduction in the intact system, nor do the contents, terpenes and pinenes (in commercial form turpentine and rosin), seem available for metabolism. (Resin pressures can be measured with small manometers.) Consequently their function is probably defensive against wind and snow damage and possibly against certain types of sucking insects. It has been reported that trees which have been induced to exude more resin suffer more than controls from 'bluestain' (caused by *Pullularia* fungus). Gum and mucilage canals behave similarly and probably have a similar function.

Xylem exudation

Root exudation

The most familiar form of exudation encountered is from xylem as root exudation following decapitation or in the intact plant as *guttation*. This phenomenon is well documented as an osmotic phenomenon and is described in Chapter 4.

Maple exudation

Less well known is the fact that xylem exudation can take place as a result of local stem pressures. These occur particularly in the genus *Acer*, of which the sugar maple is the most important, forming the basis for a small industry. Positive pressures develop in the tree trunks in the leafless condition and coincide with rising temperatures. Indeed if a twig is cut from a tree near freezing point, it is so sensitive to small temperature changes that warmed in the fingers the cut twig stump exudes sap. Trees are tapped by hammering one to four pipes directly into the tree boles near ground level. Sap collects, up to about 4 litres per tree per day. It usually contains around 4 per cent solutes, but these may increase to over 7 per cent. In industry this is boiled down to give maple syrup, a luxury food in North America.

A schematic indication of events in a single tapped maple tree are shown in Figure 10.5. As the day proceeds the temperature rises and sap is expelled, but transpiration of water curtails exudation before the temperature begins to fall by reducing the water balance of the tree. At night the roots develop a suction and so replenish their water balance. Owing to the valve-action of wood, though sap can escape, air cannot enter the moist porous walls of damaged vessels and cells in the xylem.

The mechanism is still not proved conclusively but seems to depend on the thermal expansion and contraction of gas bubbles contained within xylem fibres (Sauter, 1974). This causes a diurnal lateral flow of

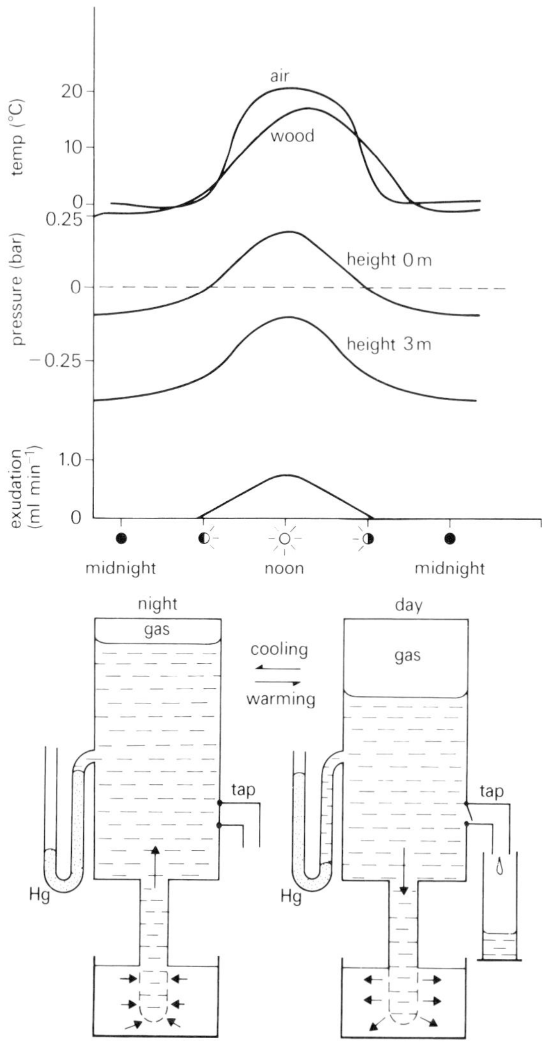

Fig. 10.5 Diagrammatic representation of Ψ_p changes in the xylem sap of a leafless *Acer* tree before bud burst. Positive pressures are greatest near ground level and seem to be generated in response to an increase in temperature. Under low temperature negative pressures are developed which restore the water balance by inducing root uptake of water.

water throughout the wood into and from the xylem vessels. Such mechanisms if operative in other plants could lead to serious errors, if relative water contents of tissues are estimated conventionally.

What then, it may be asked, is the function of this process in the intact tree? Recent evidence suggests (see Sauter, Iten and Zimmermann, 1973) that this is largely a mobilisation process in which starch and other reserves are liberated enzymically from ray cells located within the xylem as the temperature increases. The sugary sap is then pushed into the xylem and swept up by the transpiration stream to nourish the growing buds and tissues in preparation for the next growing season. Seemingly, therefore, root pressure in birches fulfils the same function as gas pressure in maple – the mobilisation and transport of solutes towards growing organs.

Pathological exudation

A quite different form of sap exudation may be encountered during forestry operations, often even in summer months. A hole bored into the xylem suddenly spurts liquid under pressure at a time when xylem sap is undoubtedly under tension. The reason for this is that the wood has been subjected to a pathogen attack producing a wet rot with gases, especially methane, as by-products. These are unable to escape so producing a local pressure in the hollow interior, and can expel the often foul-smelling decomposition fluid through a bore hole. Though this phenomenon is of importance in forestry it is not strictly physiological. It is noted here to explain what might otherwise appear to be a baffling physiological phenomenon.

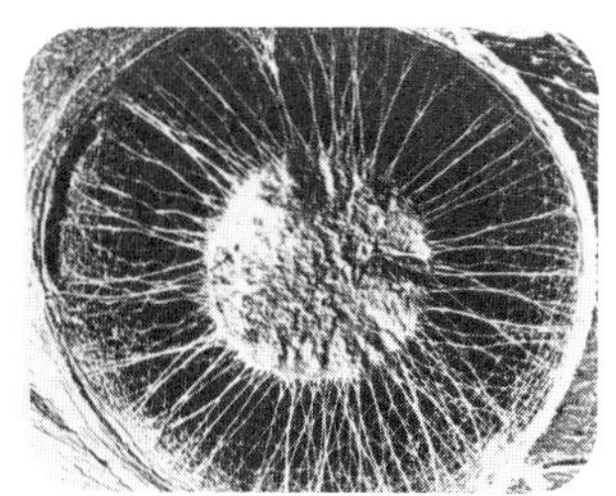

Epilogue

In concluding it is fitting to comment on the plant as a marvel of biological engineering. The constructional materials which endow plants with the capacity to resist environmental extremes and attacks by predators are themselves diverse and remarkable. Their proportions vary subtly so that plants have designs appropriate to their particular ecological environment. The great diversity of plant species indicates how many different yet viable combinations can be utilised to optimise conditions favouring growth. Present day plants represent the combination of severe evolutionary tests which have optimised light interception and also water and nutrient absorption. Nevertheless plants are protected by devices such as pit membranes and sieve plates and thanks to chemicals such as latex, resins and poisons they are by no means defenceless against potential predators.

Plants have exploited physico-chemical mechanisms such as osmosis most effectively, but sometimes enigmatically. It is commonly assumed, because the root system of a detopped plant can generate pressures like an osmometer, that osmosis must at least aid in driving the transpiration stream through a plant. This is unlikely to be correct under conditions favouring transpiration because the flow of water through the semi-permeable membranes in the root displaces the solute gradients. Under these conditions solutes accumulate outside the osmotic membranes while adjacent to the membranes inside the plant they are depleted, setting up an osmotic gradient which opposes the transpiration stream. Only when transpiration approaches zero, for example at night, do solutes accumulated by the plant promote water absorption. Nevertheless osmotic changes modify the transpiration pathway indirectly. The turgor pressures of root cells control the intimacy of soil–root contact and a shrunken root has a reduced capacity to absorb water. Similarly, transpiration from a leaf is controlled by osmotic changes in guard cells which modify the gas transport pathway. At the same time osmotic phenomena are directly involved in plant support and the patterns of plant growth and play a critical role in establishing turgor pressure gradients which transport phloem sap in bulk.

The properties of water have been utilised in a remarkable way. Land plants have an amazing capacity to maintain liquid water continuity at considerable negative pressures by stabilising their internal environment to favour cohesion. Consequently it has been possible to harness solar energy to irrigate the tissues; in contrast a positive pressure system would require the considerable expenditure of metabolic energy. Similarly, plants have exploited transport through finely porous media and utilised the surface tension of water to give a valve-like action, permitting contact with air yet preventing disruption of the conducting systems by gas.

It seems the main qualitative aspects of water transport in plants are now completely understood. Those who believe that phenomena, such as the ascent of sap in trees, are still mysterious, are simply poorly informed. This does not mean that all the minor details of the processes have been defined, for example, a great deal of work remains to be performed on ultrastructural water relations. In future the major developments lie firmly in the fields of quantitative physiology and its relation to ultrastructure, biochemistry, biophysics and ecology. Many of the measurements necessary to understand the broad quantitative aspects of water transport in plants have now been made. Without doubt the precision and facility in making measurements will improve in future as electronic instrumentation is increasingly applied to measure pressure, flow and osmotic gradients in the ecosphere and within the plant. At the same time the use of orbiting satellites utilising infrared photography has enabled preliminary diagnosis of drought and even disease in plant communities with remarkable ease when compared with land-based survey techniques.

Nevertheless it would be a mistake to imagine that all future developments lie in the fields of advanced technology. From textbooks one tends to assume a dogmatic attitude towards plants based on a single generalised plant built of standardised components. Such an appraisal can lead to dangerous misunderstandings. Plants differ from one another both structurally and physiologically. At present some of the physiological differences between plants can only be guessed from their anatomical structure; a short perusal of Haberlandt's classical *Physiological Anatomy* indicates the vast amount of background work necessary to provide the detailed picture we need of the plant community as a whole. It is also surprising how little is known about vascular organisation in many plants (even 'textbook' types), and until the last decade understanding of the vasculature of palms can only be described as abysmally neglected. Inventions such as the shuttle microscope go a considerable way to allowing us to investigate the problems in detail for the first time. Doubtless the tedium in such work will be reduced eventually by computerised robots.

Plants differ from one another physically and biochemically and the differences enable them to survive by exploiting ecologically adverse areas of the world. The stature of trees and herbs is obviously different, but underlying this are differences in physiology, for example, the deleterious effects of cavitation may be tolerable in a small herb but would be catastrophic in a large tree. Differences in drought avoidance strategy are becoming more clearly defined, but a great deal of measurement must be done before we can classify plants with confidence into those adopting one physiological stratagem of survival as distinct from another. There is good reason to believe that as we understand growth substances better we should be able not only to increase actual yields and harvesting procedures but also utilise synthetic hormones to achieve water conservation. Biochemical storage products in some plants are fats but in others starch – to what extent do these promote frost hardiness? The main sugar translocated is the disaccharide sucrose – what are the physiological implications when trisaccharides such as raffinose and stachyose are utilised instead? Photosynthesis in C_4 plants seems more efficient than in C_3 systems in terms of water-use efficiency, because the ratio of carbon fixed to water utilised is often 2:1. Thus variation in photosynthetic biochemistry seems to be advantageous for plant survival in arid and saline zones.

Nevertheless vast areas of the world are poorly productive deserts. Leaving aside the now obvious fact that many have been produced and are maintained through ignorance or incompetent management by man (e.g. through permitting grazing by goats), we must plan for times when their fertility can be increased. By studying the diverse physiology of different plants there is a reasonable chance that we can learn to identify those characters most advantageous in the various locations and improve on nature by breeding chosen plants with these qualities. Similar work in the past provided large increases in the yields of conventional agricultural crops (e.g. 'miracle rice' in India). Plant breeding is also providing uniquely useful plants with physiological defects from which much can be learned scientifically. It is also most likely that similar advances, such as those introduced by balanced fertiliser control, will increase productivity very significantly.

By learning more about plant translocation we automatically increase our capacity to control pests and pathogens by feeding the chemicals into the vascular system. It has been estimated that £60 million of damage was caused in 1970 in Britain by fungal pathogens alone. Irrigation on a cost efficiency basis can yield large returns, even in temperate regions, using existing technology. Taking a single example, the barley crop produced in Britain alone is currently about 9 million tonnes but, despite the fact that barley is reasonably drought resistant, the failure to irrigate causes a loss conservatively estimated at 5 per cent (450,000 tonnes). Crops like the potato suffer more from drought, especially in dry years. A note of caution is necessary, however, when stressing the need for greater efficiency. Natural biological systems tend to operate harmoniously with relatively minor disturbances from year to year but a narrow pursuit of efficiency from a single crop led in Ireland to the catastrophic effects of the potato famine (1845–47).

It should not be assumed that advances in plant physiology and allied sciences offer a panacea for world problems. The logical necessity to limit the human population within a framework of the earth's productive capacity has been stressed since Malthus. It says much for technological advances coupled with a blatant acceptance of human degradation and starvation that man has postponed legislation on these matters so long. The vegetation of the earth is our most important guaranteed resource, because it predominates in the transformation and storage of solar energy on which world biology depends. It is symptomatic of governmental inability to govern adequately, and so to pursue the proper long-term goals in these areas, that many of the technical skills available are neglected or badly applied (e.g. the disastrous groundnut scheme after the Second World War). The symbiosis between technologically advanced and developing countries which Britain especially could have done much to promote, has not flourished. Perhaps in future, as world needs become even more pressing, this tendency can still be improved.

In this short book I have attempted a difficult synthesis. I have tried to extend the work of others to establish a format, so that water conduction through soil plants and atmosphere can be expressed in the same conductivity units firmly based on gradients of water potential. Many of the critical parameters have not yet been measured with the required precision to convince the scientific community of their magnitudes. These often vary depending on environmental conditions, but specimen calculations have been included to indicate their probable magnitude on a comparative basis. Above all the approach I have taken is as simple as I can make it, requiring little beyond elementary algebra. I hope that in future we can avoid errors made in the past, for example when plant systems were assumed to be unrealistically simple, so that xylem conduits were expected to behave in *all* respects like simple pipes. Perhaps the last word should be from Leonardo da Vinci: 'Remember, when discoursing on the flow of water, to adduce first experience, then reason.'

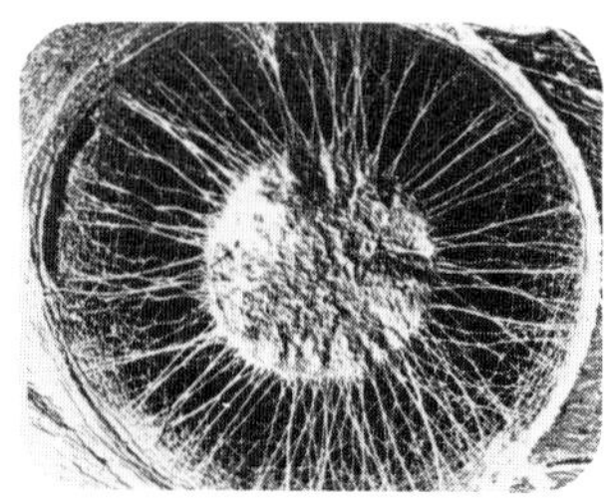

Appendices

1(a) List of symbols, parameters, dimensions and units

Symbol	Parameter	Dimensions	Typical units
A	Area of surface	L^2	m^2
A_r	Surface areas subject to radiation	L^2	m^2
A_e	Surface areas of epidermal surface	L^2	m^2
A_i	Surface areas of internal gas space	L^2	m^2
b	Non-osmotic volume in a cell	L^3	m^3
c	Concentration	ML^{-3}	kg m^{-3}
c_{sat}	Water vapour in air at saturation	ML^{-3}	kg m^{-3}
c_o	Water vapour in air below saturation	ML^{-3}	kg m^{-3}
C	Conductance (electrical)	IV^{-1}	siemens
C_p	Specific heat	$L^2T^{-2}K^{-1}$	J kg^{-1} °C^{-1}
D	Diffusion coefficient	L^2T^{-1}	m^2 s^{-1}
e	Natural logarithm	–	2.71828
E	Electrical potential difference	V	volts
g	Gravitational constant	LT^{-2}	m s^{-2}
h	Height (vertical)	L	m
I	Electrical current	A	amperes
J	Volume flow (hydraulic)	L^3T^{-1}	m^3 s^{-1}
J_v	Hydraulic flux	$L^3L^{-2}T^{-1}$	m^3 m^{-2} s^{-1} (or m s^{-1})
J_{wv}	Hydraulic flux of water vapour	$L^3L^{-2}T^{-1}$	m^3 m^{-2} s^{-1} (or m s^{-1})
ln	Natural logarithm	–	2.303 log (base 10)
L	Hydraulic conductivity	$L^3L^{-1}(ML^{-1}T^{-2})^{-1}T^{-1}$	m^2 Pa^{-1} s^{-1}
L_p	Hydraulic conductance	$L^3L^{-2}(ML^{-1}T^{-2})^{-1}T^{-1}$	m Pa^{-1} s^{-1}
M	Relative molecular mass (or weight)	M	kg
m	Molality	mol M^{-1}	mol kg^{-1}
N	Avogadro's number	–	–

P	Pressure (hydrostatic)	$ML^{-1}T^{-2}$	Pa(= N m^{-2} = bar × 10^{-5})
r	Radius	L	m
R	Resistance (electrical)	VI^{-1}	ohms
$R_{(a\,s\,or\,c)}$	Diffusive resistance to water vap.	TL^{-1}	s m^{-1}
R	Gas constant	–	8.3143 J mol^{-1} K^{-1}
r.h.	Relative humidity	–	–
RWC	Relative water content (of tissue)	–	–
S	Surface tension	$(MLT^{-2})L^{-1}$	N m^{-1}
t	Time	T	s
T	Temperature	–	K or °C
$\bar{u}$	Velocity (mean)	LT^{-1}	m s^{-1}
u_{max}	Velocity (peak mean)	LT^{-1}	m s^{-1}
V	Volume	L^3	m^3
$\bar{V}$	Partial molal volume	L^3 mol^{-1}	m^3 mol^{-1}
x	Length (or depth) of pathway	L	m
z	Valency (≡ No. osmotic particles)	–	–
Δ (delta)	Difference between two values	–	–
ε (epsilon)	Volumetric elastic modulus	$ML^{-1}T^{-2}$	–
η (eta)	Viscosity (dynamic)	$ML^{-1}T^{-1}$	kg m^{-1} s^{-1} (= N)
μ (mu)	Micro (i.e. × 10^{-6})	–	–
λ (lambda)	Thermal conductivity	–	W m^{-1} °C^{-1}
Π (pi)	Osmotic pressure ($-\Psi_s$)	$ML^{-1}T^{-2}$	Pa (= 10^{-5} bar)
ρ (rho)	Density	ML^{-3}	kg m^{-3}
σ (sigma)	Reflection coefficient	–	–
ϕ (phi)	Osmotic coefficient	–	–
Ψ (psi)	Water potential	$ML^{-1}T^{-2}$	Pa (= 10^{-5} bar)
Ψ_e	external	$ML^{-1}T^{-2}$	Pa (= 10^{-5} bar)
Ψ_g	caused by gravity	$ML^{-1}T^{-2}$	Pa (= 10^{-5} bar)
Ψ_s	caused by solutes	$ML^{-1}T^{-2}$	Pa (= 10^{-5} bar)
Ψ_m	caused by matrices	$ML^{-1}T^{-2}$	Pa (= 10^{-5} bar)
Ψ_p	caused by pressure	$ML^{-1}T^{-2}$	Pa (= 10^{-5} bar)
Ψ_t	caused thermally	$ML^{-1}T^{-2}$	Pa (= 10^{-5} bar)

1(b) The International System (SI) Units

SI bases and prefixes

Quantity	Name of	Symbol
length	metre	m
mass	kilogram	kg
time	second	s
electric current	ampere	A
thermodynamic temperature	kelvin	K
amount of substance	mole	mol
luminous intensity	candela	cd

Factor	10^{12}	10^{9}	10^{6}	10^{3}	10^{2}	10^{1}	10^{-1}
Prefix	tera	giga	mega	kilo	hecto	deca	deci
Symbol	T	G	M	k	h	da	d
Factor	10^{-2}	10^{-3}	10^{-6}	10^{-9}	10^{-12}	10^{-15}	10^{-18}
Prefix	centi	milli	micro	nano	pico	femto	atto
Symbol	c	m	μ	n	p	f	a

Named units derived from SI bases

Quantity	Name of unit	Symbol	Expression in terms of other units	Expression in terms of SI base and supplementary units
frequency	hertz	Hz		s^{-1}
force	newton	N		$m\,kg\,s^{-2}$
pressure, stress	pascal	Pa	$N\,m^{-2}$	$m^{-1}\,kg\,s^{-2}$
energy, work, quantity of heat	joule	J	N m	$m^{2}\,kg\,s^{-2}$
power, radiant flux	watt	W	$J\,s^{-1}$	$m^{2}\,kg\,s^{-3}$
quantity of electricity, electric charge	coulomb	C		s A
electric potential, potential difference, electromotive force	volt	V	$W\,A^{-1}$	$m^{2}\,kg\,s^{-3}\,A^{-1}$
capacitance	farad	F	$C\,V^{-1}$	$m^{-2}\,kg^{-1}\,s^{4}\,A^{2}$
electric resistance	ohm	Ω	$V\,A^{-1}$	$m^{2}\,kg\,s^{-3}\,A^{-2}$
conductance	siemens	S	$A\,V^{-1}$	$m^{-2}\,kg^{-1}\,s^{3}\,A^{2}$
magnetic flux	weber	Wb	V s	$m^{2}\,kg\,s^{-2}\,A^{-1}$
magnetic flux density	tesla	T	$Wb\,m^{-2}$	$kg\,s^{-2}\,A^{-1}$
inductance	henry	H	$Wb\,A^{-1}$	$m^{2}\,kg\,s^{-2}\,A^{-2}$
luminous flux	lumen	lm		cd sr
illuminance	lux	lx	$lm\,m^{-2}$	$m^{-2}\,cd\,sr$

Note

Electronic calculators have removed much tedium from calculations. It is imperative however to follow the SI rules (p. 189) and to be especially methodical in assigning powers of 10. Carelessness gives poor results which can quickly undermine confidence.

Other SI derived units

Quantity	Symbol	Expression in terms of SI base and supplementary units
area	m^2	
volume	m^3	
speed, velocity	$m\,s^{-1}$	
acceleration	$m\,s^{-2}$	
wavenumber	m^{-1}	
density, mass density	$kg\,m^{-3}$	
current density	$A\,m^{-2}$	
magnetic field strength	$A\,m^{-1}$	
concentration (of amount of substance)	$mol\,m^{-3}$	
activity (radioactive)	s^{-1}	
specific volume	$m^3\,kg^{-1}$	
luminance	$cd\,m^{-2}$	
angular velocity	$rad\,s^{-1}$	
angular acceleration	$rad\,s^{-2}$	
dynamic viscosity	$Pa\,s$	$m^{-1}\,kg\,s^{-1}$
moment of force	$N\,m$	$m^2\,kg\,s^{-2}$
surface tension	$N\,m^{-1}$	$kg\,s^{-2}$
heat flux density, irradiance	$W\,m^{-2}$	$kg\,s^{-3}$
heat capacity, entropy	$J\,K^{-1}$	$m^2\,kg\,s^{-2}\,K^{-1}$
specific heat capacity, specific entropy	$J\,kg^{-1}\,K^{-1}$	$m^2\,s^{-2}\,K^{-1}$
specific energy	$J\,Kg^{-1}$	$m^2\,s^{-2}$
thermal conductivity	$W\,m^{-1}\,K^{-1}$	$m\,kg\,s^{-3}\,K^{-1}$
energy density	$J\,m^{-3}$	$m^{-1}\,kg\,s^{-2}$
electric field strength	$V\,m^{-1}$	$m\,kg\,s^{-3}\,A^{-1}$
electric charge density	$C\,m^{-3}$	$m^{-3}\,s\,A$
electric flux density	$C\,m^{-2}$	$m^{-2}\,s\,A$
permittivity	$F\,m^{-1}$	$m^{-3}\,kg^{-1}\,s^4\,A^2$
permeability	$H\,m^{-1}$	$m\,kg\,s^{-2}\,A^{-2}$
molar energy	$J\,mol^{-1}$	$m^2\,kg\,s^{-2}\,mol^{-1}$
molar entropy, molar heat capacity	$J\,mol^{-1}\,K^{-1}$	$m^2\,kg\,s^{-2}\,K^{-1}\,mol^{-1}$
radiant intensity	$W\,sr^{-1}$	$m^2\,kg\,s^{-3}\,sr^{-1}$
radiance	$W\,m^{-2}\,sr^{-1}$	$kg\,s^{-3}\,sr^{-1}$

Rules for SI usage

1. Each quantity in a calculation must be a number times a unit. Each must be in basic SI units (e.g. m, kg, s) or units derived from them (e.g. J, N, Pa). Most tabulated data (App. 2 & 3) is in this form.
2. If SI compatible units are used appropriate factors must be introduced (e.g. bar $= 10^5$ Pa) to be consistent with basic units.
3. If doubtful over units or formulae it is useful to analyse the dimensions ensuring two sides of the equation balance.

2 Physical properties of water (Relative molecular mass=0.018016 kg

Temperature T °C	Density ρ 10^2 kg m^{-3}	Specific heat C_p 10^3 J kg^{-1} °C^{-1}	Dynamic viscosity η 10^{-4} kg m^{-1} s^{-1} (or) Pa s
0	9.9987	4.22	17.9
10	9.9973	4.19	13.1
15	–	–	–
20	9.9823	4.18	10.1
25	9.97	–	–
30	9.9568	4.18	8.0
40	9.9225	4.18	6.5
Note: 3.98 °C	10.0		

Solubilities of gases in water

Temperature 7° C	CO_2	O_2	N_2
	m^3 at STP per m^3 water at 1.013 bar		
0	1.713	0.0489	0.0239
10	1.194	0.0379	0.0196
20	0.878	0.0309	0.0164
30	0.665	0.0282	0.0138
40	0.530	0.0231	0.0118

Velocity of sound { (air STP) 331.4 m s^{-1}; water 20°C) 1,457 m s^{-1}; (dry wood) *c.* 3,500 m s^{-1}

3 Physical properties of water vapour

Temperature T °C	Density (H_2O) saturated 10^{-3} kg m^{-3}	Vapour pressure saturated (H_2O) 10^{-8} N m^{-2} (10^{-3} bar)	Vapour pressure saturated H_2O mm Hg
0	4.85	6.1	4.6
5	6.80		6.52
10	9.401	12.3	9.2
15	12.83		12.8
20	17.30	23.3	17.5
25	23.05	31.7	23.8
30	30.38	42.4	31.8
40		73.7	55.3

Miscellaneous

Avogadro's number
(number of molecules per mole) 6.022×10^{23}

Gravitational acceleration $g = 9.807$ ms^{-2}

Partial molal volume V 10^{-6} m^3 mol^{-1}	Surface tension S 10^{-3} N m^{-1} (or) J M^{-2}	Thermal conductivity λ W m^{-1} °C^{-1}
18.018	75.6	0.55
18.021	74.2	0.58
18.032	–	–
18.048	72.8	0.60
18.056	–	–
18.094	71.2	0.62
18.157	69.6	0.63

Note on use of tables
To find the true value of any quantity from these tables the number in the column is multiplied by its heading, e.g. the density of water at 25°C = 9.97×10^2 kg m^{-3}.

Concentration water vapour	in air (saturated)	Thermal conductivity	Heat of H_2O vaporisation	Diffusion coeff. H_2O vap.
C_{wv} 10^{-3} kg m^{-3}	C_{wv} mol m^{-3}	λ 10^{-2} W m^{-1} °C^{-1}	10^4 J mol^{-1}	D_{H_2O} 10^6 m^2 s^{-1}
4.85	0.269		4.49	22.6
6.8	0.378			23.3
9.41	0.522		4.45	24.1
12.8	0.712			24.9
17.3	0.96		4.41	25.7
23.0	1.28		4.39	26.5
30.4	1.69		4.33	27.3
51.1	2.84		4.31	

Atmospheric pressure 1.013×10^5 N m^{-2} (= 1.013 bar)
Kinematic viscosity (water) $V = \eta/\rho$ 1.51×10^{-5} m^2 s^{-1}
Relative molecular mass (H_2O) 18.016×10^{-3} kg
1 coulomb = 1 ampere per second ≡ 6.0×10^{18} electrons charge

4 Pressure systems

Comparative pressure systems encountered in everyday life in familiar and SI units for comparison with pressures encountered in plants (from various sources)

System	Familiar units	SI units Pa ($N\ m^{-2}$)	Bar
Human blood pressure (systolic)	110 mm Hg	1.5×10^4	0.15
Human blood pressure (diastolic)	70 mm Hg	9.0×10^3	0.09
Human blood pressure (venous)	10 mm Hg	1.0×10^3	0.01
Air pressure in car tyres (typical)	30 lb sq inch	2.0×10^5	2.0
Air pressure in racing cycle tyres	120 lb sq inch	8.0×10^5	8.0
Water pressure mains (max. approx.)	99 lb sq inch	6.0×10^5	6.0
Water pressure hot water (2-storey house)	30 lb sq inch	2.0×10^5	2.0
Air diving cylinder (typical)	150 atm	1.52×10^7	152.0
Atmospheric pressure	1 atm	1.013×10^5	1.013

5 Theoretical estimates of the tensile strength of water

1. Energy content (Nobel's method)
It is assumed that at 25°C water has only about 80 per cent of the hydrogen bonds remaining. Knowing that 9.6 kcal mol^{-1} represents the strength of bonding in water at 0°C, the fraction remaining at 20°C is $0.8 \times 9.6 = 7.7$ kcal $mol^{-1} = 0.43 \times 10^3$ kcal $kg^{-1} = 0.443 \times 10^6$ kcal m^{-3}, using standard conversion factors to convert moles of water to cubic metres. Now 1 kcal = 4,180 J = 4,180 N m^{-1} and 1 bar = 10^5 N m^{-2} (i.e. Pa).

So $0.443 \times 10^6 \times 4{,}180 \times 10^{-5} = 18{,}500$ bar, the tensile strength of water. The weakness of this method lies in the assumption of bond energies because vaporisation does not necessarily involve total fracture of all bonds.

2. Thermal equivalent method
At 100°C the tensile strength of water is zero because water molecules vaporise. Each reduction of 1°C increases the tensile strength by 81.6 bar (see Fig. 6.10).

$$\left.\begin{array}{l}\text{So at } 25°C = 75 \times 81.6 = 6{,}120 \text{ bar} \\ \text{and at } 0°C = 100 \times 81.6 = 8{,}160 \text{ bar}\end{array}\right\} \text{Pressures holding molecules together}$$

Values derived from this method are reasonable and based on experiment but extrapolation may not be justified over the range 0° to 100°C. (A change occurs in the structure of water at +4°C, for example.)

6 Relative humidity from wet and dry bulb hygrometry °C

Values are approx. for an airspeed over 3 ms^{-1} and a barometric pressure of 742.7 mm Hg (from Bull. U.S. Weath. Bur. 1071)

Ambient temperature °C	Temperature differential below ambient °C								
	1	2	3	4	5	8	10	13	14
0	81	64	46	29	13	–	–	–	–
5	86	72	58	45	33	–	–	–	–
10	88	77	66	55	44	15	–	–	–
15	90	80	71	61	53	27	13	–	–
20	91	83	74	66	59	37	24	12	–
25	92	84	77	70	63	44	33	22	12
30	93	86	79	73	67	50	39	30	21
40	94	88	82	77	72	57	48	40	33

7 Osmolality and *RT*

An *osmole* corresponds with an ideal gas in that 1 mol dissolved in water exerts 1 bar osmotic pressure when dissolved in *RT* litres. Correspondingly if 1 mol is dissolved in 1 litre (1 kg) of water it will generate *RT* bar osmotic pressure ($R = 8.3143$ J mol^{-1} K^{-1})

Temperature °C	0	5	10	15	20	25	30
RT litre bar mol^{-1} K^{-1}	22.71	23.11	23.53	23.94	24.37	24.79	25.19
Temperature K	273	278	383	288	293	278	303

8 Relationship between water potential of water vapour of different relative humidities at frequently encountered temperatures

Relative humidity	$-\Psi_{wv}$ at different temperatures °C (bar)				
	10	15	20	25	30
100	0	0	0	0	0
99.5	6.54	6.66	6.77	6.88	6.98
99	13.12	13.34	13.57	13.79	13.99
98	26.38	26.82	27.27	27.72	28.13
95	66.9	68.11	69.23	70.38	71.42
90	137.5	139.9	142.2	144.5	146.6
80	291.3	296.3	301.1	306.1	310.6
70	465.6	473.6	481.4	489.4	496.5
50	905.0	920.4	935.5	951.1	965.0
30	1,571.9	1,598.7	1,625.0	1,652.0	1,676.2
10	3,006.3	3,057.5	3,107.9	3,159.5	3,205.8
$\frac{RT}{\bar{V}_w}$	1,305.6	1,327.9	1,349.7	1,372.7	1,392.2

The above was calculated from the formula

$$-\Psi_{wv} = \frac{RT}{\bar{V}_w} \ln\left(\frac{\text{per cent r.h.}}{100}\right)$$

Thus for 50 per cent r.h.

$$-\Psi_{wv} \text{ at } 20°\text{C} = \frac{0.083141 \times 293}{18.048 \times 10^{-3}} \ln(0.5) = 935.58 \text{ bar.}$$

Towards 0 per cent relative humidity the calculation becomes progressively less valid. Note that all Ψ_{wv} values are negative.

9 Calculation of osmotic pressure of a pure solution

If an osmotic solution with a water potential Ψ_s is placed in an osmometer at equilibrium, it produces pressure so raising the water potential Ψ_p by pressure to cancel out Ψ_s. Then

$$\Psi = 0 = \Psi_p + \Psi_s \qquad \text{and } \Psi_p = -\Psi_s$$

Ψ_p is the osmotic pressure of the solution (symbolised as Π for the general case). Providing the solution is 'ideal' it accords with Van't Hoff equations derived from the gas laws thus:

$$\Psi_p = -\Psi_s = \frac{RT}{\bar{V}}\frac{Mm}{1{,}000}.$$

Many dilute solutions behave in this ideal manner, but more concentrated solutions do not. Also the ions in electrolytes behave as independent osmotically active particles. To allow for this behaviour an osmotic coefficient and valency must be included for precise information.

$$\Psi_p = \frac{RT}{\bar{V}} \frac{M\,m\,z\,\phi}{1{,}000}$$

Component	Symbol	Dimensions	Units
Osmotic pressure	$\Psi_p = -\Psi_s$	$ML^{-1}T$	Pa (or $N\,m^{-2}$ or $J\,m^{-3}$)
Gas constant	R	$ML^2T^{-2}\,mol^{-10}\,K^{-1}$	$J\,mol^{-10}$ K^{-1}
Absolute temperature	T	K	K
Relative molecular mass solvent	M	$mol\,M^{-1}$	$mol\,kg^{-1}$
Molal concentration of solute	m	$M\,mol^{-1}$	$kg\,mol^{-1}$
Valency of solute	z	–	–
Osmotic coefficient	ϕ	–	–
Water potential	Ψ	$ML^{-1}T^{-2}$	Pa (or $N\,m^{-2}$ or $J\,m^{-3}$)
Partial molal volume	$\bar{V}$	$L^3\,mol^{-1}$	$m^3\,mol^{-1}$

Thus $ML^{-1}T^{-2} = \dfrac{ML^2T^{-2}\,mol^{-1}\,K^{-1}}{L^3\,mol^{-1}}\,K\,\dfrac{M}{mol}\,\dfrac{mol}{M}$

Example

Thus for 0.6 molal solution of NaCl at 25°C

$$\Psi_p = \frac{8.3143 \times 298}{0.018056 \times 10^{-6}} \times 18.0016 \times 0.6 \times 10^{-6} \times 2 \times 0.9230 \times 10^{-5} = 27.36\ \text{bar}$$

10 Osmotic coefficient ϕ and osmotic pressures Π (corr. to 2 decimal places) of molal solutions of sucrose and common electrolytes (derived from Robinson and Stokes 1959)

It is advisable to derive intermediate values from a large scale graph for the selected area. It should be noted that both the partial molal volume of water and osmotic coefficients used in these derivations vary with temperature in a manner not necessarily proportional to absolute temperature K

Molality	**Sucrose**		**Sodium chloride**		**Potassium chloride**		**Calcium chloride**	
mol litre^{-1}		bar 25°C		bar 25°C		bar 25°C		bar 25°C
m	ϕ	Π	ϕ	Π	ϕ	Π	ϕ	Π
0.1	1.008	2.49	0.9324	4.62	0.9266	4.58	0.854	6.33
0.2	1.017	5.03	0.9245	9.14	0.9130	9.03	0.862	12.79
0.3	1.024	9.20	0.9215	13.66	0.9063	13.44	0.876	19.50
0.4	1.033	10.21	0.9203	18.20	0.9017	17.83	0.894	26.52
0.5	1.041	12.87	0.9209	22.76	0.8989	22.22	0.917	34.00
0.6	1.050	15.57	0.9230	27.38	0.8976	26.63	0.940	41.83
0.7	1.060	18.34	0.9257	29.67	0.8970	31.04	0.963	49.99
0.8	1.068	21.12	0.9288	36.74	0.8970	35.48	0.988	58.62
0.9	1.079	24.01	0.9320	41.47	0.8971	39.92	1.017	67.88
1.0	1.088	26.70	0.9355	46.25	0.8974	44.37	1.046	77.57
1.2	1.108	32.87	0.9428	55.94	0.8986	53.31	1.107	98.52
1.4	1.129	39.07	0.9513	65.85	0.9010	62.36	1.171	121.58
1.6	1.150	45.48	0.9616	76.07	0.9042	71.53	1.237	146.78
1.8	1.169	52.02	0.9723	86.53	0.9081	80.81	1.305	174.20
2.0	1.189	57.79	0.9833	97.23	0.9124	90.22	1.376	204.09
2.5	1.240	76.63					1.568	290.71
3.0	1.288	95.52	1.0453	155.04	0.9367	138.93	1.779	395.80
3.5	1.334	115.42					1.981	514.20
4.0	1.375	135.96	1.1158	220.66	0.9647	190.78	2.182	647.28
4.5	1.414	157.29					2.383	795.27
5.0	1.450	179.22	1.1916	294.57			2.574	954.45
5.5	1.482	201.49					2.743	1118.83
6.0	1.511	224.11	1.2706	376.92			2.891	1286.40

11(a) Approximate osmotic potentials Ψ (bar) of solutions of given molal[1] and molar[2] concentration at 20°C showing the difference between molal and molar solutions as the concentration is increased

Note that salts with two ions have double the osmotic potential of non-electrolytes when dilute but the effect is compensated for by the volume of sucrose at higher molarities.

Concentration	Molal sucrose	Molar sucrose	Molar KCl	Molar NaCl
0.1	−2.6	−2.6	−4.5	−4.2
0.2	−5.0	−5.2	−8.7	−8.3
0.3	−7.5	−8.0	−12.4	−12.3
0.4	−10.0	−11.0	−16.5	−16.4
0.5	−12.6	−14.1	−20.3	−20.7
0.6	−15.2	−17.6	−23.8	−25.4
0.7	−17.9	−21.2	−27.9	−29.3
0.8	−20.6	−25.2	−31.8	−33.8
0.9	−23.4	−29.3	−35.4	−38.1
1.0	−26.3	−34.1	−38.7	−42.6

1. A molal solution is 1 gram relative molecular mass *plus* 1 kg of solvent (water) i.e. $1\,\mathrm{mol\,kg^{-1}}$.
2. A molar solution contains 1 gram relative molecular mass *in* 1 litre of solution i.e. $1\,\mathrm{mol\,dm^{-3}}$.

11(b) Derivation of molal solutions by dilution of a 1 molal solution

Molal solutions cannot be diluted volumetrically like molar solutions. Dilution must be performed by taking a given weight of molal solution in a tare on a top loading balance and adding the required quantity of water. These quantities to make over a litre of solution can be calculated from the formula below, all quantities being given in grams.

Molality x Solution required	$=$ (1,000 + relative molecular mass solute)x Molal solution (g)	$+$ $(1{,}000 - 1{,}000x)$ Water (g)

12(a) Refractive index (at 20°C) and osmotic pressure Π of solutions of convenient osmotica expressed in terms of percentage concentration (grams solute per 100 grams *solution*)

Osmotic pressures can be converted to osmoles (1 osmole = 24.3746 bar at 20°C). Intermediate values can be determined conveniently by plotting graphs of refractive index ($\Delta RI \times 10^4$ gives the difference from pure water) against osmotic pressure (derived from *Handbook of Physics and Chemistry*).

Aqueous concentration g solute per 100 g solution	Sodium chloride		Calcium chloride		Glycerol $C_3H_8O_3$		D-Mannitol $C_6H_{14}O_6$		Sucrose $C_{12}H_{22}O_{11}$	
	ΔRI x10^4	OP bar	ΔRI x10^4	OP bar	ΔRI x10^4	OP bar	ΔRI x10^4	OP bar	ΔRI x10^4	OP bar
1.0	18	7.78	24	5.79	12	2.36	15	1.34	14	0.73
2.0	35	15.55	48	11.53	23	5.39	29	2.72	29	1.46
5.0	88	39.92	121	30.74	58	14.14	73	7.07	73	3.80
10.0	175	86.02	245	76.78	118	30.44	147	15.04	148	8.19
16.0	282	155.75	400	160.87	191	53.65	–	–	243	14.38
24.0	428	291.59	621	331.49	294	91.89	–	–	376	24.72

12(b) Osmotic pressures (in bar) and refractive indices of sucrose solutions at 20°C

Concentrations are expressed in gram moles per litre volume of solution: first decimals are on vertical axis. Second places of decimals are shown on the horizontal axis. Refractive index is shown as the difference from pure water $\times 10^4$.

Concentrations volume molar	$\Delta RI \times 10^4$	Second decimals 0.00	0.01	0.02	0.03	0.04	0.05
0.0	0	0.00	0.26	0.54	0.80	1.07	1.34
0.1	49	2.67	2.95	3.21	3.47	3.75	4.01
0.2	98	5.36	5.64	5.94	6.22	6.50	6.79
0.3	147	8.23	8.52	8.82	9.12	9.41	9.70
0.4	196	11.24	11.55	11.85	12.26	12.56	12.87
0.5	245	14.49	14.79	15.20	15.50	15.80	16.21
0.6	294	18.03	18.34	18.74	19.15	19.45	19.85
0.7	343	21.78	22.18	22.59	23.00	23.40	23.70
0.8	391	25.83	26.34	26.74	27.15	27.55	27.96
0.9	439	30.09	30.59	31.10	31.50	32.01	32.52
1.0	486	35.05	35.56	36.16	36.67	31.18	37.68

Refractometers designs

Bench refractometers have a wider range but need a temperature controlled bath for high accuracy. Hand refractometers fit the pocket and are extremely convenient. The best designs have inbuilt temperature compensation over 0–400 ΔRI e.g. Goldberg AOTS from: American Optical, Scientific Instruments Division, Buffalo, NY, 14215, USA, or British AO, Instrument Group, 870 Yeovil Road, Slough, Bucks, UK, or Atago Refractometers from ChemLab Instruments Ltd, Hornminster House, 129 Upminster Road, Hornchurch, Essex RM11 3XJ.

0.06	0.07	0.08	0.09
1.61	1.87	2.14	2.41
4.27	4.54	4.81	5.08
7.07	7.36	7.65	7.94
10.00	10.33	10.64	10.94
13.17	13.47	13.88	14.18
16.61	16.92	17.32	17.63
20.26	20.67	20.97	21.37
24.11	24.62	25.02	25.43
28.36	28.77	29.17	29.68
33.02	33.53	34.04	34.54
38.19	38.70	39.30	39.81

13 Alternative units for hydraulic conductivity L in soils

Some soil scientists use a unit for hydraulic conductivity different from the units used throughout this book ($m^2\ bar^{-1}\ s^{-1}$). This arises from the fact that they often use columns of water to drive hydraulic fluxes by gravitation. Hence instead of bars the equivalent units would be metres height, where 1,000 cm $\approx$ 1 bar depending on density (and hence temperature). Consequently dimensions *seem* quite different, e.g. $cm\ h^{-1}$ or $cm\ s^{-}$ These values can be converted easily into SI units however, thus: $10\ cm\ s^{-1} = 1\ m^2\ bar^{-1}\ s^{-1}$ (because 1 cm vertical water column over 1 cm horizontal distance equals $0.1\ bar\ m^{-1}$).

14 Notes on thermodynamic terminology and usage

(a) Thermodynamic terminology and Onsager coefficients
It may be wondered why certain terminology has been used in this book – the use of subscripts may seem confusing, since they are descriptive rather than algebraic terms. So far as possible the symbols and units have been chosen to equate with those commonly in use today to analyse more complex systems than those dealt with in the text, so that the reader may graduate to more complex systems using irreversible thermodynamics relatively easily. SI units were scheduled to be standard usage by December 1977.

Thermodynamics was originally concerned with equilibrium conditions, i.e. steady state or 'thermostatic'. Advances made by Onsager (1931) led to the analysis of irreversible (non-steady-state) thermodynamics dealing with flows affected by pressure, diffusional and electrical potentials. Equations can be conjugated as follows:

Total	**Flow components** (conductivity coeff. × driving potential (1–3))		
Flow system =	(1) Pressure +	(2) Diffusion +	(3) Electrical
Water J_v ($m^3\ m^{-2}\ s^{-1}$)	L_P pressure grad.	L_{PD} diffusion grad.	L_{PE} electrical grad.
Solutes J_D ($mol\ m^{-2}\ s^{-1}$)	L_{DP} pressure grad.	L_D diffusion grad.	L_{DE} electrical grad.
Electrons J_E (A)	L_{EP} pressure grad.	L_{ED} diffusion grad.	L_E electrical grad.

It will be seen that the above pressure, diffusion and electrical potentials have more familiar names, e.g. the electrical potential for water flow is electro-osmosis; for electrical potentials affecting solutes is electrophoresis. At first sight there appear to be nine coefficients of L. However, those indicated with double letter symbols are identical thus:

$L_{PD} = L_{DP}$, being 'cross-coefficients'.

The six coefficients are measurable only be isolating them. For pressure, one could reduce the effect of diffusion and electricity to units of pressure. Thus by manipulation all components are given in terms of L_p using the usual symbolism:

$$\text{Overall } J_v = L_p(\Delta\Psi_p + \Delta\Psi_s) + L_p[1-\sigma]\Delta\Psi_s - L_p\frac{[P_E I]}{K}.$$

To find the pressure term L_p the diffusion and electrical terms are reduced to zero, so that

$$L_p = \frac{J_v}{(\Delta\Psi_p + \Delta\Psi_s)}$$

Also, since $K = \frac{I}{E}, \frac{C_i - C_o}{2}(1-\sigma) = \frac{J_s}{J_v}$,

the values of K, the specific conductance, and σ, the reflection coefficient, can be determined. In this manner it is possible to combine results to deduce J_v, as in the equation above, involving pressure, diffusional and electrical components operating together.

(b) Use of differentials

To simplify calculations, a common assumption made in the text is that the gradient of a given driving component is a linear function (which means that if plotted on a graph it would be a straight line). This is increasingly legitimate the smaller the distance involved, but over longer path lengths a driving component may be non-linear and follow a more complex function, like the sigmoid 'front' of solute molecules driven by diffusion. In many expressions a more precise terminology would be introduced by replacing

$\frac{\Delta c}{x}$ or $\frac{\Delta c}{\Delta x}$ with $\frac{dc}{dx}$, and similarly $\frac{\Delta\Psi_p}{\Delta h}$ with $\frac{dp}{dh}$, etc.

In most instances the simple assumption is sufficiently accurate when compared with the precision of measurements but possible exceptions should be borne in mind.

(c) The Nernst equation – usage

In many situations in biology, such as studies of ion uptake by cells, or exudation from decapitated root systems, the sap contains a large array of cations and anions (and also solutes such as sucrose, which are not ionised). Strictly speaking the Nernst equation is only applicable to an individual ionic species. The equation should also be considered strictly in terms of activity coefficients rather than the concentration of ions we have used. In fact the rate of activity coefficients approximately equals the ratio of concentrations when the internal and external concentrations are equal but ceases to be true when the concentrations are very dissimilar. For these reasons the Nernst equation can only be used as a guide in deciding the active or passive transport of an ion and further information is necessary for confirmation.

(d) Water potential terminology

The normal thermodynamic definition of a potential (μ) is based on energy per mole ($J\,mol^{-1}$). The thermodynamic potential of water potential Ψ retains the name 'potential' to emphasise the energy concept which can be made, but it has the dimensions of pressure to accommodate the measurements most conveniently made on plants. According to this definition

$$\Psi = \frac{\mu - \mu_o}{\bar{V}}$$

where the water potential in units of pressure is the difference between the chemical potential (μ) of a given sample of water with that of pure water (μ_o) per volume of 1 mol of water ($\bar{V}$). This terminology is rigorous and meets the requirements of an interdisciplinary approach (see Slatyer, 1967) but is not essential for most practical or theoretical plant physiology.

15 A note on computation of gaseous resistances and conductances

Diffusive gaseous resistances

In series $R_{total} = R_1 + R_2$ (Where R_1 may be at leaf surface and R_2 may be gaseous phase)

In parallel $\frac{L}{R_{total}} = \frac{1}{R_1} + \frac{1}{R_2}$ (Where R_1 may be cuticular resistance and R_2 may be stomatal resistance)

Thus $R_{total} = \frac{R_1 R_2}{R_1 + R_2}$

Diffusive gaseous conductances

Let gaseous diffusive conductance $\frac{1}{R} = G$

In series $\frac{1}{G_{total}} = \frac{1}{G_1} + \frac{1}{G_2}$

so $G_{total} = \frac{G_1 G_2}{L_1 + L_2}$

In parallel $G_{total} = G_1 + G_2$

16 Silicone rubber impressions of leaf surfaces

The method is essentially that of Sampson (1961) using 'Silflo' silicone rubber and 'Silflex' catalyst, supplied by J. & S. Davis Ltd, London. The rubber catalyst mixture is rapidly placed on the leaf surface and allowed to set. This primary replica is placed directly on a glass slide freshly coated with clear nail varnish and allowed to set for 30 min before removing the primary. The secondary impression can be examined directly with a $40\times$ objective. If oil immersion is used a coverslip is needed to protect the impression. Care must be taken to ensure the correct 'throat' is measured.

17 Useful water-relations definitions

1. *Water potential* Ψ_o is the standard water potential of pure free water at atmospheric pressure and the same temperature as a system under investigation. It is arbitrarily set at 0 bar or Pa.
2. *Water potential* Ψ is the total potential comprising the sum of contributory potential e.g. Ψ_p and Ψ_s. All water potentials (as below) are measured at equilibrium and related to the reference water potential Ψ_0 at the same temperature and pressure.
3. *Pressure potential* Ψ_p is the increase in water potential above zero through positive pressurisation or the reduction below zero by negative pressurisation (suction). It is positive or negative in sign.
4. *Osmotic* (*or solute*) *potential* Ψ_s is the *reduction* in the water potential by the addition of one or more solutes. It is invariably negative in sign.
5. *Matric potential* Ψ_m is the *reduction* in the water potential by the presence of solid, often finely divided, surfaces. It is invariably negative in sign.
6. *Osmotic pressure* Π is the positive hydrostatic pressure generated by a solution of given osmotic potential Ψ_s when it is placed in an osmometer in the presence of pure water under isothermal conditions at equilibrium ($\Pi = \Psi_s$ numerically, but differs in sign).

7. *Turgor pressure* Ψ_p is the hydrostatic pressure developed in a cell which is usually positive in living cells but negative in xylem conduits. It depends on the net water potential tending to drive water into the cell i.e. Ψ_s within opposed by $\Psi_p + \Psi_s$ outside the cell. When the external Ψ is zero, if the membrane is perfectly semi-permeable, the turgor pressure Ψ_p at equilibrium is identical with Π the osmotic pressure of the cell sap.
8. (*a*) *Osmole.* An osmole is the osmotic pressure of 1 mol of an ideal solute dissolved in 1 kg.
   (*b*) An osmole is alternatively the osmotic pressure of a solution at 0°C with a freezing point depression of 1.86°C. *Note:* 1 Osmole equals 1,000 mosmoles. Thus from (*a*) an osmole has an osmotic pressure of 22.71 bar at 0°C or an osmole has an osmotic pressure of 24.37 bar at 20°C using the conversion factor ($\times 273 + 20/273$) for temp. adjustment. Note that the osmole is *not* a precise expression of osmotic pressure because: (i) the osmotic coefficient varies with temperature; (ii) in a multicomponent solution intersolute interactions can occur which decrease precision (see Dick, 1966); (iii) 'ideal solute' assumptions only apply with accuracy to very dilute solutions.
9. *Summation of water potentials.* Water potentials are *added* together, but the fact that potentials can be negative has caused confusion

   $$\Psi = \Psi_p + \Psi_s + \Psi_m$$

   Example (*1*): Find Ψ in a cell where Ψ_p is +10 bar, Ψ_s is −9 bar and Ψ_m is −1 bar.
   $\Psi = (+10) + (-9) + (-1) = 0$ bar, i.e. at equilibrium in pure water.

   Example (*2*): Find Ψ in a xylem conduit given $\Psi_p = -5$ and $\Psi_s = -1$ bar.
   $\Psi = (-5) + (-1) = \underline{-6}$ bar, i.e. the presence of solutes augments the negative pressure of −5 bar in lowering Ψ.
10. *Alternative water potential terminology.* Several attempts have been made to simplify the equation above for work on cells to remove the brackets above.
   (*a*) *Pressure units.* Instead of working in water potentials equivalent pressure units have been used *with a negative sign* to deduce total water potential. Thus: $\Psi = P - \Pi$ where (P = hydrostatic pressure, Π = osmotic pressure). This arrangement is legitimate providing the nature of the components is clearly understood.

   Its disadvantages seem to me as follows:

   (1) A 'potential' equals the difference between two pressures which does not seem consistent. Ψ in these terms could be called, more honestly, a *water pressure.*

   (2) The advantage of water potential terminology is that it does focus attention on the *solvent*, water. The measurement of a pressure P loses this desirable emphasis which might apply equally say to ε, the volumetric elastic modulus (see Ch. 3).

   (3) An osmometer is a device for measuring the negative Ψ_s by generation of a positive hydrostatic pressure ($\Psi_p = \Pi$). In a similar way a pressure bomb also measures negative sap pressures by means of a balancing positive gas pressure. Both *positive* pressures measure implicit *negative* water potentials (either Ψ_s or Ψ_p) but the pressures themselves are of opposite sign. Thus despite the apparent simplicity of the relation $\Psi = P - \Pi$, it really should be more correctly written $\Psi = P + (-\Pi)$.

   (*b*) *Potential units.* Some have sought to use 10(a) terminology calling pressures potentials and retaining the negative sign. *This must be deplored* (even if the units are properly defined) because it leads to unconventional use of Π which *should* be osmotic pressure, but which instead becomes $-\Pi$.

   (*c*) *Energy units.* The thermodynamic definition of water is based on the difference between the chemical potential (μ) of water in a system with that of pure water (μ_0) at STP per volume of 1 mol of water. In this way though water potential is considered in terms of energy per mole it has the dimensions of pressures as explained in Appendix 14*d*.

(*d*) *A note on pressure terminology*. It is possible to define pressure in terms of the Van't Hoff gas law equation:

$$P = \frac{nRT}{\bar{V}}$$

where the symbols have the same meaning as defined previously. It is apparent that when n (the number of particles) becomes zero, the pressure P is zero also (i.e. perfect vacuum). It is impossible for P to become negative *in gases*.

In this book pressure has been considered in terms of force per unit area. In this form it *can* be applied readily to cohesion of liquids (as in the Berthelot tube, see Ch. 2) or to solids (such as the force per unit cross-sectional area in a wire which is supporting a load under tension). Pressure in this sense is vectorial, and to be manifest requires some rigid container or

18 Conversion factors

Length		(inch $= 0.0254$ m)
		1 cm $= 10^{-2}$ m
Area		(sq. inch $= 6.452 \times 10^{-4}$ m^2)
		1 cm$^2 = 10^{-4}$ m^2
Volume		1 cm$^3 = 10^{-6}$ m^3
		(cub inch $= 1.639 \times 10^{-5}$ m^3)
		litre $= 10^{-3}$ m^3 ($= 1$ dm^3)
Density	kg (water)	$\simeq 10^{-3}$ m^3
	ml	$= 10^{-6}$ m^3 ($\cong 1$ g water)
	μl	$= 10^{-9}$ m^3 ($\cong 1$ mg water)
	pl	$= 10^{-12}$ m^3 ($\cong 1$ μg water)
Mass	1 g	$= 10^{-3}$ kg
Concentrations	mol m^{-3}	$= 10^{-6}$ mol cm^3
	mol	$=$ kg (rel. mol mass)$^{-1}$ m^{-3}
Pressure	1 atmosphere	$= 1.013$ bar
	1 dyn cm^{-2}	$= 0.1$ N m^{-2}
	1 inch water (4°C)	$= 249.1$ N m^{-2}
	1 mm Hg (0°C)	$= 133.3$ N m^{-2}
	1 bar	$= 0.987$ atm
	1 bar	$= 10^5$ J m^{-3}
	1 bar	$= 10^2$ J kg^{-1}
	1 bar	$= 10^5$ N m^{-2}
	1 bar	$= 10^5$ Pa
	1 bar	$= 10^2$ J kg^{-1} (water)
	1 bar	$= 10^6$ dyn cm^{-2}
Force	1 dyn	$= 10^{-5}$ N
	1 kg F	$= 9.81$ N
Viscosity	1 poise	$= 10^{-2}$ m s^{-1}

Hydraulic conductance	1 m bar^{-1} s^{-1}	$=10^2$ cm bar^{-1} s^{-1}
Hydraulic conductivity	1 m^2 bar^{-1} s^{-1}	$=10^4$ cm^2 bar^{-1} s^{-1}
	1 m^2 bar^{-1} s^{-1}	$=10$ cm s^{-1} (H_2O column)
Flow	cm s^{-1}	$=10^{-2}$ m s^{-1}
Time	Hour	$=3.6\times10^3$ s
	Day	$=8.64\times10^4$ s
	Year	$=3.15\times10^7$ s
Miscellaneous		
Length	Angstrom	$=10^{-10}$ m
Pressure	Torr	$=1$ mm Hg column $=1.33\times10^{-4}$ bar (at sea level)
Power of radiation	Watt W	$=1$ J s^{-1} ($=10^7$ erg s^{-1})
	1,000 Wm^{-2}	$=1.43$ cal cm^{-2} min^{-1}

Useful data

Gas laws For water $\bar{V}=18.048\times10^{-6}$ m mol^{-1} at 20°C
$R=83.141\times10^{-6}$ m^3 bar K^{-1}
$RT/\bar{V}=1{,}260$ bar 0°C; 1,306 bar 10°C; 1,350 bar 20°C; 1,372 bar 25°C.

Osmotic pressure: Freezing point $\Delta 1$°C $\equiv 22.71$ bar 0°C
$\equiv 24.37$ bar at 20°C

Nernst equation Faraday $F=96{,}487$ J mol^{-1} V^{-1}
$RT/F=25.3$ mV at 20°C $=25.7$ mV at 25°C
$2.303\ RT/F=58.2$ mV at 20°C $=59.2$ mV at 25°C

Avogadro's number $=6.022\times10^{23}$ mol^{-1}

Boltzmann's constant $=8.617\times10^{-5}$ eV/molecule K
$=1.380\times10^{-3}$ J K^{-1}

Gravitational acceleration g $=9.806$ m s^{-2} (at 45° latitude)

Solar constant $=1{,}390$ W m^{-2}

19 Properties of aqueous solutions and gases

Density, refractivity and viscosity of aqueous solutions. All are relative to water at 20°C.

Substance dissolved in water	Property at 20°C relative to water at 20°C		
	Density	Refractivity ($\times10^4$)	Viscosity
10% by wt.	$\Delta\rho$	ΔRI ($\times10^4$)	$\Delta\eta$
Calcium chloride	1.0854	245	1.316
Potassium chloride	1.0652	136	0.986
Sodium chloride	1.0726	175	1.191
Methanol	0.9833	24	1.326
Ethanol	0.9836	65	1.498
Glycerol	1.0233	118	1.188
D-Mannitol	1.0357	147	1.353
Glucose	1.0393	147	1.327
Sucrose	1.0400	148	1.333

Typical surface tension values for pure liquids and aqueous solutions

Substance, pure or in water	**Concentration** % by wt	**Surface tension** $S 10^3$ N m^{-1}	
		10°C	20°C
Ethanol	100	23.61	22.75
Glycerol	100	–	63.40
Water	100	74.22	72.90
Glycerol	10	–	72.90
Sodium chloride	10	–	76.00

Excess pressure developed by free bubbles and bubbles confined in conduits as a result of surface tension *S* of pure water (from Eq. [2.2])

Radius of bubble or conduit at narrowest bore	**Excess gas pressure developed (bar)**		
	0°C	10°C	20°C
10×10^{-9}	151.20	148.40	145.60
$10^2 \times 10^{-9}$	15.10	14.80	14.60
$10^3 \times 10^{-9}$	1.51	1.48	1.46
$10^4 \times 10^{-9}$	0.15	0.15	0.15

Diffusion coefficients *D* of several solutes in water at 25°C

Substance	**Conc.**	**Ionised solute**	**Substance**	**Non-ionised solutes**
Potassium chloride	0.001 *M*	1.964	Mannitol (v. dilute)	0.682
Sodium chloride	0.001 *M*	1.585	Glucose (v. dilute)	0.673
Calcium chloride	0.001 *M*	1.263	Sucrose (v. dilute)	0.523
Potassium chloride	0.1 *M*	1.844	Methanol (v. dilute)	1.370
Sodium chloride	0.1 *M*	1.483	Ethanol (v. dilute)	1.100
Calcium chloride	0.1 *M*	1.110	Propanol (v. dilute)	0.980

Typical values for the surface tension *S* of pure liquids and aqueous solutions

Liquid or Substance in water (10% by wt.)	Surface tension *S* N m$^{-1} \times 10^{-3}$	
	10°C	20°C
Ethanol	23.61	22.75
Glycerol	–	63.40
Glycerol in water	–	72.90
Water	74.22	72.75
Sodium chloride in water	–	76.00

Diffusion coefficients *D* of gases at different temperatures
(For water vapour see App. 3)

Gases mixed or in air	Diffusion coefficient D		
	10°C	20°C	30°C
Oxygen	0.25	0.26	0.27
Carbon dioxide	0.18	0.19	0.19
Nitrogen	0.18	0.19	0.20

20 Relative atomic mass and symbols of common elements

Element	Symbol	Relative atomic mass	Element	Symbol	Relative atomic mass
Hydrogen	H	1.008	Sulphur	S	32.07
Carbon	C	12.011	Chloride	Cl	35.46
Nitrogen	N	14.008	Potassium	K	39.10
Oxygen	O	16.000	Calcium	Ca	40.08
Sodium	Na	22.99	Iron	Fe	55.85
Magnesium	Mg	24.31	Iodine	I	126.90
Phosphorus	P	30.98	Lead	Pb	207.21

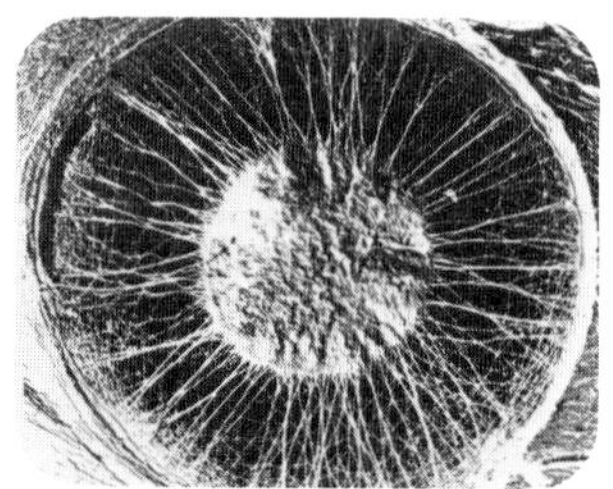

Problems

1. What is the water potential Ψ of a solution which freezes at $-1.32\,^{\circ}C$: (*a*) at $0\,^{\circ}C$; (*b*) at $20\,^{\circ}C$?
2. Radioactive tracer with a long half-life is applied to a tissue at a point A. At intervals the radioactivity above background is measured at point B to see if the tracer is carried by a flowing stream of liquid or if it merely spreads by diffusion. Calculate the time (seconds) the radioactivity is expected to reach 37 per cent at B of the level at A if the distance A–B is: (*a*) 1 millimetre; (*b*) 0.5 metre.
3. What is the water potential Ψ of an air sample the humidity of which is 78 per cent r.h. at: (*a*) $20\,^{\circ}C$; (*b*) $25\,^{\circ}C$?
4. The flow of sap J_v through a narrow conduit is at the rapid rate of $10\,m\,s^{-1}$. Given that the conduit radius is $10\,\mu m$, the sap density ρ is $10.05\times10^{2}\,kg\,m^{-3}$ and its viscosity n is $15.0\times10^{-4}\,kg\,m^{-1}\,s^{-1}$ calculate: (*a*) from the Reynolds number if flow is turbulent or laminar, and, (*b*) deduce the flow rate J_v at which laminar flow is expected to give way to turbulence.
5. Calculate the osmotic pressure Π at $25\,^{\circ}C$ of: (*a*) a 0.5 molal aqueous solution of sucrose (osmotic coefficient $\phi = 1.041$); (*b*) a 1.2 molal solution of potassium chloride ($\phi = 0.8986$), and; (*c*) a 2.5 molal solution of calcium chloride ($\phi = 1.568$). Assume $\bar{V} = 18.056\times10^{-3}\,m^{3}$ at $25\,^{\circ}C$ and $R = 83.143\times10^{-6}\,mol\,bar^{-1}\,K^{-1}$.
6. The hydrostatic pressure within xylem at the base of a tall vine in the early spring is 1.5 bar. Calculate the pressure at a height of 20 m: (*a*) if the sap is virtually pure water, and; (*b*) if the sap is sugary with a density ρ of $1{,}050\,kg\,m^{-3}$. Assume that the gravitational constant $g = 9.81\,m\,s^{-2}$.
7. A rigid, porous, highly-wettable tube, sealed at both ends, is filled with water and then subjected to an internal negative pressure (i.e. liquid is placed under tension). Cavitation occurs at $20\,^{\circ}C$ when the internal pressure is about -5 bar. The largest pores in the walls revealed by electron microscopic tests have radii of 20 nm. Does cavitation seem to occur because atmospheric air is sucked through these pores, or from some other cause? The surface tension of water S is $7.3\times10^{-2}\,N\,m^{-1}$ at $20\,^{\circ}C$.
8. An osmometer, filled with glycerol with an osmotic potential Ψ_s of -12 bar, is immersed in pure water. The intervening membrane has an osmotic coefficient σ

for glycerol of 0.75. (*a*) Calculate the maximum hydrostatic pressure P developed by the osmometer. Will it persist? (*b*) Calculate the *maximum* hydrostatic pressure which might develop if the same osmometer contains propanol also ($\Psi_s = -6$ bar) in addition to the glycerol (σ for propanol is 0.5). (*c*) Why might the hydrostatic pressure be lower than the *maximum* expected and; (*d*) How could the hydrostatic pressure be maintained for long periods?

9. The xylem conduits of a tree have cavitated and become embolised with air. The conduit radii vary from 0.5 to 20 μm. Calculate the *maximum* radius of conduits which we expect to refill with sap when the xylem sap tension is reduced for several hours by rainfall to (*a*) 1.0 bar (*b*) 0.3 bar. The surface tension of the xylem sap was 74.2×10^{-3} N m^{-1}.

10. A cell with a volume of 1×10^{-5} m^3 reaches equilibrium with an external water potential of -6 bar. In this condition the cell vacuolar sap has an osmotic potential Ψ_s of -15 bar. When the external water potential increases to -2 bar the cell volume is increased by 30 per cent when equilibrated. (*a*) What is the initial turgor pressure Ψ_p of the cell? (*b*) What is the new Ψ_s of the vacuolar sap after the change; (*c*) What is the new Ψ_p exactly? (*d*) What is the volumetric elastic modulus ε of the cell under these conditions?

11. An algal cell is immersed in a fluid at 25°C with the following ionic composition, K^+:0.1 m*M*; Na^+:1.0 m*M*, Cl^-:1.3 m*M* and H PO_4^-:0.1 m*M*. The vacuolar sap electrical potential E_{obs} is -87 mV relative to the external medium and has the following composition K^+:40 m*M*; Na^+:17 m*M*; Cl^-:1.3 m*M* and H PO_4^-:0.1 m*M*. The vacuolar sap electrical potential E_{obs} is -87 mV relative to the external medium and has the following composition K^+:40 m*M*; Na^+:17 m*M*; Cl^-:38 m*M* and $H_2PO_4^-$:1.5 m*M*. Calculate the expected Nernst potentials E_N for each ion at equilibrium and so, assuming the Nernst equation is applicable, if ions are being actively accumulated or expelled by the cell.

12. Solutes are stirred into the hydroponic medium of a decapitated root system until exudation from the cut stump ceases. The freezing point of the root exudate is -0.09°C and that of the hydroponic fluid with added solute is -0.23°C. (*a*) Calculate the osmotic pressure apparently generated within the root. (*b*) Explain the discrepancy and suggest why a hydrostatic pressure of 1.7 bar had to be applied to the root system in the original hydroponic medium (freezing point 0.04°C) to stop exudation.

13. A decapitated root system in water is sealed in a pressure bomb with the stem stump protruding. Pressure P is applied and the rate of sap exudation is measured. Initially additional pressure P is applied. P of 0.5 bar induces 2 ml volume of sap V to exude in 10 minutes with an osmotic potential Ψ_s of 0.15 bar. Later when P is 1.0 bar V is 5.5 ml in 10 minutes and Ψ_s becomes 0.05 bar. Calculate the hydraulic conductance of the root system (*a*) initially when $P = 0.5$ bar; (*b*) when P is 1.0 bar. (*c*) How might the values of Ψ_s and V change if P is zero. (*d*) Are the conductance changes explained by the changes in Ψ_s (Root area $A = 12 \times 10^{-4}$ m^{-2}.)

14. The flow of sap (hydraulic flux J_v) up a tree 10 m tall is on average 1 m per hour. The pressure difference driving flow is 1 bar overall. The wooden trunk cross-section is 1 m^2. (*a*) Calculate the hydraulic conductance L_p of the wooden trunk. (*b*) Calculate the hydraulic conductivity L of the wooden trunk.

15. Calculate the hydraulic conductivity L of two uniform conduits at 20°C. Their radii are 1×10^{-6} m and 1×10^{-3} m. (The viscosity of water η is 10.1×10^{-4} Pa s at 20°C.)

16. A pressure drives a flow of water through a conduit 10 cm long with a radius of 10^{-4} at a rate of 1 cm^3 per 100 s at 20°C. (Viscosity $\eta = 10.1 \times 10^{-4}$ Pa s at 20°C.) Calculate (*a*) the hydraulic flux in m s^{-1} and (*b*) $\Delta\Psi_p$ in bar.

17. A vessel in a ring-porous tree is 20 m long with a mean radius of 1 mm. The rate of sap flow (or hydraulic flux) J_v is 10 m s^{-1}. Calculate the pressure gradient (cf.

press. diff. ΔP) driving the flow at 20°C. (Viscosity of the sap η was 10.1×10^{-4} Pa s at 20°C.)

18. A pressure gradient of 0.1 bar m^{-1} drives water through four conduits of different bore in a vine stem. Calculate the volume of sap (viscosity $\eta = 10.1 \times 10^{-4}$ Pa s) flowing each second through each of the four tubes given their radii are (*a*) 1×10^{-3} m; (*b*) 1×10^{-4} m; (*c*) 1×10^{-5} m; and (*d*) 1×10^{-6} m.

19. In a xylem conduit 1 m long and cross-sectional area 400 $(\mu m)^2$ a pressure gradient of 0.1 bar m^{-1} produces a flow of water of 0.02 ml (assume 1 cm^3 = 1 ml) in 3,600 s. Calculate (*a*) the rate of sap flow, i.e. hydraulic flux J_v per hour; (*b*) the hydraulic conductance (L_p) of the conduit; and (*c*) what is the conduit radius r in μm?

20. The transpiration of an excised leaf was found to be steady at 3×10^{-8} m s^{-1} by measuring the uptake with a balance. The relative humidity of its boundary layer (by the osmotic droplet technique) was 89 per cent r.h. at 25°C (by thermocouple). When the leaf was tested in a pressure bomb the pressure of incipient exudation was 7.2 bar and Ψ_s of the sap, expelled by additional pressure was 0.8 bar. (*a*) What is the water potential Ψ of the leaf xylem; (*b*) What is the water potential difference $\Delta\Psi$ across the leaf tissues if at 25°C; and (*c*) if the leaf had been 26.1°C. (*d*) Calculate the hydraulic conductance L_p of the leaf tissues.

21. (*a*) Calculate the total gaseous resistance R_{total} of a leaf resistance R_L and the gas layer resistance R_G in which $R_L = 80$ s m^{-1} and $R_G = 400$ s m^{-1}. (*b*) Calculate R_{total} of the cuticular resistance R_c and stomatal resistance R_s of a leaf surface given $R_c = 5{,}400$ s m^{-1} and $R_s = 185$ s m^{-1}. (*c*) Calculate gaseous conductances G for the above values for (*a*) and (*b*).

22. Calculate the mean boundary gas layer thickness (x) over a leaf surface over which a wind flows if (*a*) this distance (l) is 20 mm and the wind velocity V is 32 m s^{-1}; or (*b*) if l is 500 mm and V is 50 m s^{-1}. Assume that the empirical formula $x = 0.4 \times 10^{-2} (1/V)^{1/2}$ is applicable.

23. A population of stomata from the underside of a leaf is examined and found to be very uniform. The pores (measured in sections) are 10 μm deep and are 11 μm long and have a frequency of 150 per mm^2 by the silicone imprint method. Calculate the stomatal gaseous resistance R_s assuming (*a*) that the pore is a cylinder with diameter 8 μm by Brown and Escombe formula; (*b*) that the pore is a diamond in cross-section with a maximum width of 5 μm; and (*c*) if the 'diamond width' narrows to 2 μm. (The diffusion coefficient D of water vapour in air is 25.7×10^{-6} m s^{-1} at the ambient 20°C.)

24. Calculate the hydraulic conductance L_p of a stomatal pore of which R_s is 216.3 s m^{-1} when the hydraulic conductivity of air above the pore (95 per cent r.h. at 20°C) is 2.8×10^{-4} m^2 bar^{-1} s^{-1} (see Fig. 6.9: D of water vapour = 25.7×10^{-6} m s^{-1}).

25. The boundary layer of air immediately surrounding a plant is 50 per cent r.h. and at 22°C. Transpiration measured by a potometer is steady at 12×10^{-5} kg m^{-2} s^{-1} from the leaf surface 500 μm thick. The hydraulic conductivity of the air is 2.3×10^{-14} m^2 bar^{-1} s^{-1}. Calculate the water potential of the outermost tissues of the transpiring plant: (*a*) at the same temperature; or (*b*) at 21°C. (*c*) What tissues are likely to be involved in this water loss?

26. A leaf in an oil bomb develops a xylem tension of 14 bar. It then absorbs 0.5 ml water via the petiole after which it can only develop a xylem tension over a period of time of 12 bar. The final osmotic potential Ψ_s of the sap, determined by a freezing point osmometer, is −15 bar. What is (*a*) the turgor pressure Ψ_p at which growth stops; (*b*) the initial Ψ_s of the leaf sap; and (*c*) its volume?

27. The turgor pressure Ψ_p of a *Ricinus* sieve tube is 11 bar and the osmotic potential Ψ_s of the sap is −20 bar. What is the water potential of an adjacent phloem parenchyma cell (*a*) when the cell reaches equilibrium with the turgor sieve tube; (*b*) when the sieve tube has been punctured so that Ψ_p falls to zero?

28. The phloem system of a 10 m tall tree contains 5×10^6 sieve tubes each of 10 μm radius. The turgor pressure Ψ_p at the tree top is 15 bar and 11 bar at the base. The sugary phloem sap has a viscosity η of 10.1×10^{-4} kg m^{-1} s^{-1} (or Pa s) and a density of ρ of 11.0×10^2 kg m^{-3}. Calculate (*a*) the pressure difference $\Delta\Psi$ and pressure gradient (assuming that $\rho = 10^3$ kg m^{-3}) driving sap flow? (Remember gravity!) Determine (*b*) the speed of sap flow J_v; (*c*) the hydraulic conductivity L of each sieve tube; and (*d*) the volume of sap V in litres of sap flowing each day assuming the Poiseuille equation is applicable. Finally, (*e*) how much dry matter (in kg) is translocated per day?

29. A bark tongue is cut from a tree but left attached above. In 30 days the dry matter in the bark flap is increased near its tip by 4 g. Assuming that this is represented by starch which has been transported in the form of sucrose (10 g in 100 ml water) via the phloem, what is the maximum amount of water which might be released (*a*) if the bark fresh weight was unchanged; (*b*) if the fresh weight increased by 2 g?

30. The xylem sap at the top of a 30 m tall maple tree has a pressure potential Ψ_p of -5 bar and an osmotic potential Ψ_s of -2 bar by night. (*a*) What is the water potential of an adjacent xylem parenchymatous cell at equilibrium with the sap by day when Ψ_s is still -2 bar but Ψ_p has become -2 bar by day; (*b*) Would a bore hole at the base of the tree exude (i) by night and (ii) by day; (*c*) If exudate were collected via the bore hole what might be its Ψ_s?

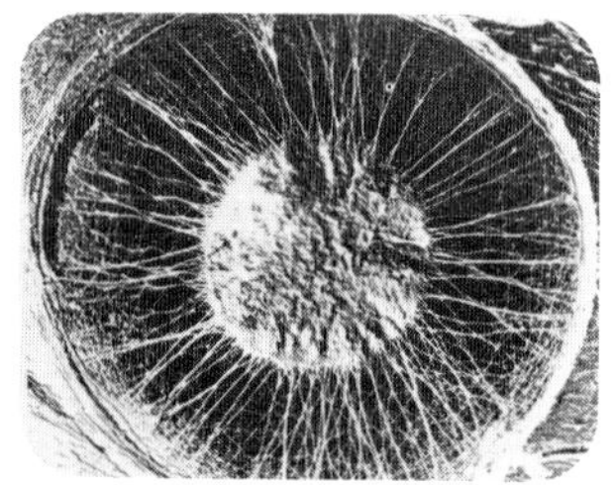

Answers to selected problems

Guidance is provided by reference to equations (Eq.) appendices (App.) and figures (Fig.)

1. (Eq. [1.2]) (*a*) $\Pi = 15.88$ bar (*b*) $\Pi = 17.05$ bar

2. (Eq. [1.12]) (*a*) $t = 371.5$ s $= 6$ min 11.5 s (*b*) $t = 92, 867, 776$ s $= 496$ years

3. (Eq. [1.3]) (*a*) $\Psi = 335.4$ bar (*b*) $\Psi = 341.1$ bar

4. (Eq. [1.9]) (*a*) Rc$=26.6$ which is much less than 2,000 so flow is laminar.
(*b*) For Re$\approx 1{,}000$ $J_v = 149.3\,\mathrm{m\,s^{-1}}$

5. (App. 9 and 10) (*a*) $\Pi = 12.87$ bar (*b*) $\Pi = 53.31$ bar (*c*) $\Pi = 290.71$ bar

6. (Eq. [2.1]) (*a*) $P = 1.5 - 1.96 = -0.46$ bar (*b*) $P = -0.56$ bar

7. (Eq. [2.2]) $20\,\mu$m pores withstand a pressure diff. of 73 bar. Dirt or internal bubbles must be presumed. (Bubbles are 292 nm radius max.)

8. (Eq. [2.6]) (*a*) $\Pi = 9$ bar max. This will decline through leakage.
(*b*) $\Pi_{max} = 12$ bar
(*c*) Π will be less than 12 if each solute P_{max} fails to coincide.
(*d*) P maintained if $\sigma = 1$ for a solute or by active solute transport.

9. (Eq. [2.2]) (*a*) $r = 1.5\,\mu$m (*b*) $r = 4.95\,\mu$m

10. (*a*) $\Psi_p = 9$ bar initially (*b*) $\Psi_s = 11.54$ bar
(*c*) $\Psi_p = 5.53$ bar (*d*) $\varepsilon = 13.3$ bar (Eq. [3.4])

11. E_N for $K^+ = -67$ mV ion accumulated; E_N for $Na^+ = +14$ mV ion expelled
E_N for $Cl^- = 0$ m V ion neutral; E_N for $HPO_4^- = -52$ mV ion expelled (Eq. [3.6])

12. (*a*) $\Psi_s = 1.09$ bar (exudate)$=2.77$ bar (medium). $\Delta\Psi_s = 1.68$ bar caused by root
(*b*) $\Psi_s = 0.48$ bar (medium). $P = 1.7$ bar so $\Psi = (-1.7) + (-0.48) = -2.18$ bar
Pressure Ψ_s of solutes at osmotic membrane was 2.18 bar or more but solutes were removed *en route* so that apparent $\Psi_s = 1.09$ bar (exudate).

13. (*a*) $L_p = 4.2 \times 10^{-5}\,\mathrm{m\,bar^{-1}\,s^{-1}}$ when $P = 0.5$ bar
(*b*) $L_p = 7.3 \times 10^{-5}\,\mathrm{m\,bar^{-1}\,s^{-1}}$ when $P = 1.0$ bar

Apparently L_p is about doubled when P is doubled. Note P and Ψ_s drive water in the same direction.

14. (Eq. [1.5]) (*a*) $L_p = 2.78 \times 10^{-4}\,\mathrm{m\,bar^{-1}\,s^{-1}}$ ($= 2.78 \times 10^{-9}\,\mathrm{m\,Pa^{-1}\,s^{-1}}$)
(Eq. [1.6]) (*b*) $L = 2.78 \times 10^{-3}\,\mathrm{m^2\,bar^{-1}\,s^{-1}}$ ($= 2.78 \times 10^{-8}\,\mathrm{m^2\,Pa^{-1}\,s^{-1}}$)

15. (Eq. [1.7]) (*a*) $L = 1.23 \times 10^{-5}\,m^2\,bar^{-1}\,s^{-1}$ ($= 1.23 \times 10^{-10}\,m^2\,Pa^{-1}\,s^{-1}$)
(Eq. [1.7]) (*b*) $L = 1.23 \times 10\,m^2\,bar^{-1}\,s^{-1}$ ($= 1.23 \times 10^{-4}\,m^2\,Pa^{-1}\,s^{-1}$)

16. (Eq. [1.6]) (*a*) $J_v = 0.32\,m\,s^{-1}$ (*b*) $\Delta\Psi_p = 0.26$ bar ($= 2.6 \times 10^4$ Pa)

17. (Eq. [1.7]) $\Delta\Psi_p = 16.16$ bar ($= 1.6 \times 10^6$ Pa) over 20 m.
$\Delta\Psi x^{-1} = 0.808\,bar\,m^{-1}$

18. (Eq. [1.7]) (*a*) $Vt^{-1} = 3.89\,ml\,s^{-1}$ ($= 3.89\,cm^3\,s^{-1}$)
(*b*) $Vt^{-1} = 3.89 \times 10^{-4}\,ml\,s^{-1}$
(*c*) $Vt^{-1} = 3.89\,ml\,s^{-1} \times 10^{-8}\,ml\,s^{-1}$
(*d*) $Vt^{-1} = 3.89 \times 10^{-12}\,ml\,s^{-1}$

19. (Eqs. [1.6] and [1.7]) (*a*) $J_v = 0.00139\,m\,s^{-1}$ ($= 1.39\,mm\,s^{-1} = 5.0\,m\,h^{-1}$)
(Eq. [1.5]) (*b*) $L_p = 0.00139\,m\,bar^{-1}\,s^{-1}$
($A = \Pi r^2$) (*c*) $r = 11.28\,\mu m$

20. (Eqs. [1.1] and [1.3]) (*a*) $\Psi_{xylem} = 8.0$ bar (*b*) $\Delta\Psi = 152$ bar
(Eq. [1.5]) (*c*) $\Delta\Psi = 152 + (1.1 \times 81.6) = 241.8$ bar
(Fig. 6.10) (*d*) (i) $L_p = 1.97 \times 10^{-10}\,m\,bar^{-1}\,s^{-1}$
(ii) $L_p = 1.24 \times 10^{-10}\,m\,bar^{-1}\,s^{-1}$

21. (App. 15) (*a*) $R_{total} = 480\,s\,m^{-1}$ (*b*) $R_{total} = 178.9\,s\,m^{-1}$
(*c*) (i) $G = 21 \times 10^{-3}\,m\,s^{-1}$
(ii) $G = 5.6 \times 10^{-3}\,m\,s^{-1}$

22. (Eq. [6.8]) (*a*) $x = 1 \times 10^{-4}$ m ($= 0.1$ mm) (*b*) $x = 0.4 \times 10^{-3}$ m ($= 0.4$ mm)

23. (Eq. [7.1]) (*a*) $R_s = 72\,s\,m^{-1}$; (*b*) $R_s = 88\,s\,m^{-1}$; (*c*) $R_s = 220\,s\,m^{-1}$

24. (Eq. [6.10]) $L_p = 5.04 \times 10^{-7}\,m\,bar^{-1}\,s^{-1}$ ($= 5.04 \times 10^{-12}\,m\,Pa^{-1}\,s^{-1}$)

25. (Eqs. [1.5] and App. 8) (*a*) $\Psi_L = 682$ bar (*b*) $\Psi_L = 600.4$ bar
(*c*) Cell walls of stomata, adjacent mesophyll cells and epidermal cells.

26. (Eq. [1.1]) (*a*) $\Psi_p = 3$ bar when growth stops
(*b*) $\Psi_p = 3$ bar so $\Psi_s = -11$ bar
(*c*) $V_{total} = 1.375$ ml (i.e. $V:11$ as $V + 0.5:15$)

27. (Eq. [1.1]) (*a*) $\Psi = -9$ bar
(*b*) $\Psi = -20$ bar assuming no influx of water to maintain Ψ_p. Water influx would raise Ψ towards zero.

28. (Eqs. [1.1] and [2.1]) (*a*) $\Psi_p = 5$ bar
(Eq. [1.7]) (*b*) $J_v = 6.19 \times 10^{-4}\,m\,s^{-1}$ ($= 0.619\,min\,s^{-1}$)
$= 2.22\,m\,h^{-1}$)
(Eqs. [1.6] and [1.7]) (*c*) $L = 1.23 \times 10^{-3}\,m^2\,bar^{-1}\,s^{-1}$
(Eq. [1.6]) (*d*) $Vt^{-1} = 0.084\,m^3\,day^{-1}$ ($= 84$ litres day^{-1})
(*e*) Dry matter $= 8.4\,kg\,day^{-1}$

29. (*a*) $2(C_6H_{10}O_5)_n$ starch, rel. mol. mass $324.3 \equiv 1(C_{12}H_{22}O_{11})$ sucrose rel. mol. mass 342.3
So 4 g starch $\equiv$ 4.22 g sucrose $\equiv$ 42.2 ml water
(*b*) If weight increased 2 g only 40.2 ml water could be exuded.

30. (Eq. [1.1]) (*a*) $\Psi = -7$ bar at 30 m by night
$\Psi = -4$ bar at 30 m by day
(*b*) (i) No, Ψ_p is negative (ii) Yes, Ψ_p is positive
(*c*) $\Psi_s = -2$ bar assuming Ψ_s is uniform throughout tree trunk.

Selected bibliography

Some references given in tabulated format have been excluded. Texts recommended for general reading and referred to in the text are shown with asterisks*; those without specific references are shown by daggers†.

ACEVEDO, E., HSIAO, T. C. and HENDERSON, D. W. (1971) *Pl. Physiol.*, **48,** 631–6.

*ANDERSON, W. P. (1976) In *Encyl. Pl. Physiol.* vol. 2B, eds. Lüttge, U. and Pitman, M. G. Springer, Berlin & New York.

ANDERSON, P. and HIGINBOTHAM, N. (1975) *J. Exp. Bot.*, **26,** 533–5.

ASKENASY, E. (1895) *Verb. naturwiss.-med.* Ver., **5,** Heidelberg.

†BANNISTER, P. (1976) *Introduction to Physiological Ecology.* Blackwell, Oxford.

*BARRS, H. D. (1969) In *Water Deficits and Plant Growth*, vol. 1. Academic Press, London.

BARRS, H. D. and WEATHERLEY, P. E. (1962) *Aust. J. Biol. Sci.*, **15,** 413–28.

BEAMENT, J. W. L. (1964) *Adv. Insect. Physiol.*, **2,** 67–129.

BEARDSELL, M. F. and COHEN, D. (1974) In Mech. regl. plant growth, ed. Bielski *et al. Bull. Roy. Soc. N. Zealand*, **12,** 411.

BERTHELOT, M. (1850) *Ann. de Phys. et de Chim.*, **30,** 232.

BOWLING, D. J. F. (1965) *Nature* (*Lond.*), **206,** 317–18.

*BOWLING, D. J. F. (1973) *Ion Transport in Plants*, ed. Anderson, W. P. Academic Press, London.

*BOWLING, D. J. F., (1976) *Uptake of Ions by Plant Roots.* Chapman and Hall, London.

BOWLING, D. J. F., MACKLON, A. E. S. and SPANSWICK, R. M. (1966) *J. Exp. Bot.*, **17,** 410–16.

BOYER, J. (1968) *Pl. Physiol.*, **43,** 1056–62.

BOYER, J. (1970) *Pl. Physiol.*, **46,** 233–5.

*BRIGGS, L. J. (1950) *J. appl. Phys.*, **21,** 721–3.

BROWN, H. and ESCOMBE, F. (1900) *Phil. Trans. R. Soc.*, **193,** 223.

BUIJS, K. and CHOPPIN, G. R. (1963) *J. chem. Phys.*, **39,** 2035.

BURSTROM, H. G. (1948) *Physiol. Plantarum.*, **1,** 57–64.

BUTTERY, B. R. and BOATMAN, S. G. (1966) *J. Exp. Bot.*, **17**, 283–96.
CLEMENTS, R. W. (1967) *Naval Stores Rev.*, **77**, 4–9.
COIMBRA-FILHO, A. F. and MITTERMEIER, R. A. (1976) *Nature (Lond.)*, **262**, 630.
COWAN, I. R. (1972) *Planta (Berl.)*, **106**, 185–219.
*CRAFTS, A. S., CURRIER, H. B. and STOCKING, C. R. (1949) Water in the physiology of plants, *Chron. Bot.*, Waltham, Mass.
*DAINTY, J. (1976) In *Transport in Plants II. A Cells*, ed. Lüttge, U. and Pitman, M. G. Springer-Verlag, Berlin, Heidelberg, New York.
DAINTY, J. and GINZBURG, B. Z. (1964) *Biochem. Biophys. Acta.* **79**, 102–11.
DARWIN, F. (1898) *Phil. Trans. R. Soc.*, **B 190**, 531–621.
DARWIN, F. and PERTZ, D. F. M. (1911) *Proc. Roy. Soc.*, **B 84**, 136.
DAUBENMIRE, R. F. (1959) *Plants and Environment*, 422.
DE SANTO, A., ALFANI, A. and DE LUCA, P. (1976) *Ann. Bot. N.S.*, **40**, 391–4.
*DICK, D. A. T. (1966) *Cell Water.* Butterworths, Washington, USA.
DIMOND, A. E. (1965) *Pl. Physiol.*, **40**, 119–31.
*DIXON, H. H. (1914) *Transpiration and the Ascent of Sap in Plants.* Macmillan, London.
DIXON, H. H. and BALL, N. G. (1922) *Nature (Lond.)* **109**, 236–7.
DIXON, H. H. and BARLEE, J. S. (1940) *Scient. Proc. R. Dublin Soc.*, **22**, 211.
DIXON, H. H. and JOLY, J. (1896) *Roy. Soc. (Lond.) Phil Trans.* **B 186**, 563–76.
DREW, D. H. (1966) *J. hort. Sci.* **41**, 103–14.
EATON, F. M. (1942) *J. Agr. Res.*, **64**, 357–99.
EDWARDS, M. and MEIDNER, H. (1975) *Nature (Lond.)*, **253**, 114–15.
EICKE, R. (1954) *Ber. dtsch. bot. Ges.*, **67**, 213–17.
EKERN, P. C. (1965) *Pl. Physiol.*, **40**, 736–9.
*EVANS, G. C. (1972) *The Quantitative Analysis of Plant Growth.* Blackwell, Oxford.
FISCHER, R. A. (1968) *Science*, **160**, 784.
*GAFF, D. F. and HALLAM, N. D. (1974) In Mechanisms of regulation of plant growth, ed. Bielski *et al. Bull. Roy. Soc. New Zealand.*, **12.**
*GIBBS, R. D. (1958) In *The Physiology of Forest Trees*, pp. 43–69. Ed. Thimann, K. V. Ronald Press, New York.
*GINDEL, I. (1973) *A New Ecophysiological Approach to Forest Water Relationships in Arid Climates.* Junk, The Hague.
GODLESWKI, E. (1884) *Jahrb. f. Wiss. Bot.*, **15**, 569.
GREEN, P. B. (1968) *Pl. Physiol.*, **43**, 1169–84.
GREEN, P. B., ERIKSON, R. O. and BUGGY, J. (1971) *Pl. Physiol.*, **47**, 423–30.
GREENIDGE, K. N. H. (1952) *Amer. J. Bot.*, **39**, 570–3.
GROSSENBACHER (1939) *Ann. J. Bot.*, **26**, 107–9.
GUNNING, B. E. S. and PATE, J. S. (1969) *Protoplasma*, **68**, 107–33.
*GUNNING, B. E. S. and PATE, J. S. (1974) Transfer cells. In *Dynamic Aspects of Plant Ultrastructure.* Ed. Robards. McGraw-Hill, London.
GUTNECHT, J. (1968) *Biochem. Biophys. Acta*, **163**, 20–9.
*HABERLANDT, G. (1914) *Physiological Plant Anatomy.* (Transl. Drummond, M.) 4th edn. Macmillan, London.
*HALES, S. (1727) *Vegetable Staticks.* Reprinted, 1961. Oldbourne, London.
HALL, S. M. and MILBURN, J. A. (1973) *Planta (Berl.)*, **109**, 1–10.
HAMMEL, H. T. (1968) *Pl. Physiol.*, **43**, 1042–8.
*HAMMEL, H. T. and SCHOLANDER, P. F. (1976) *Osmosis and Tensile Solvent*. Springer, Berlin and New York.
HARTSOCK, T. L. and NOBEL, P. S. (1976) *Nature (Lond.)*, **262**, 574–6.
HEATH, O. V. S. (1938) *New Phytol.*, **37**, 385–95.
HEATH, O. V. S. (1949) *New Phytol.*, **48**, 186–211.
*HEATH, O. V. S. (1975) *Stomata.* Oxford U.P., London.
HEINE, R. W. (1970) *Ann. Bot.*, **34**, 1019–24.

HEINICKE, A. I. and CHILDERS, W. F. (1936) *Proc. Amer. Soc. Hort.*, **33,** 155–9.
HELKVIST, J., RICHARDS, G. P. and JARVIS, P. G. (1974) *J. Appl. Ecol.*, **11,** 637–68.
HOLMGREN, P., JARVIS, P. G. and JARVIS, M. S. (1965) *Physiol. Plantarum*, **18,** 557–73.
*HOUSE, C. R. (1974) *Water Transport in Cells and Tissues.* Arnold, London.
HOUSE, C. R. and FINDLAY, N. (1966) *J. Exp. Bot.*, **17,** 627–40.
HSIAO, T. C. (1973) *Ann. Rev. Pl. Physiol.*, **24,** 519–70.
*HUBER, B. (1956) In *Encyclop. Pl. Physiol.*, vol. 3. ed. Ruhland. Springer-Verlag, Berlin.
HYGEN, G. (1963) In *Water Stress in Plants*, pp. 89–98. Ed. Slavik. Junk, The Hague.
ILJIN, W. S. (1957) *Ann. Rev. Pl. Physiol.*, **8,** 257–74.
IMAMURA, S. (1943) *Jap. J. Bot.*, **12,** 251–347.
IVANOFF, L. (1928) *Ber. dtsch. bot. Ges.*, **46,** 306.
JARVIS, P. G. (1975) Water transfer in plants. In *Heat and Mass Transfer in Vegetation* (Symp. Dbrovick). Ed. de Vries Pub. Scripta, Washington D.C.
JENNINGS, D. H. (Ed.) (1977) *Integration of Activity in the Higher Plant.* Cambridge U.P., Cambridge.
JONES, H. G. (1976) *J. Appl. Ecol.* **13,** 605–22.
JORDAN, W. R. (1970) *J. Agron.* **62,** 699–701.
KAMIYA, N. and TAZAWA, M. (1956) *Protoplasma*, **46,** 394–422.
*KOZLOWSKI, T. T. (1969) *Water Deficits and Plant Growth*, vol. 2. *Plant Water Consumption and Response.* Academic Press, London.
KRAMER, P. J. (1949) *Plant and Soil Relationships.* McGraw-Hill, New York.
*KRAMER, P. J. (1969) *Plant and Soil Water Relationships.* McGraw-Hill, New York.
†LANGE, O. L., KAPPEN, L. and SCHULZE, E.-D. (Eds.) (1976). *Water and Plant Life.* Springer-Verlag, Berlin, Heidelberg and New York.
LEVITT, J. (1957) *Pl. Physiol.*, **32,** 248–51.
*LEVITT, J. (1972) *Response of Plants to Environmental Stresses.* Academic Press, New York and London.
†LEYTON, L. (1975) *Fluid Behaviour in Biological Systems.* Clarendon Press, Oxford.
LLOYD, F. E. (1908) *Publ. Carnegie Inst. Wash.*, **82,** 1.
LOFTFIELD, J. V. G. (1921) *Publ. Carnegie Inst. Wash.*, **314,** 1–104.
†LUTTGE, U. and PITMAN, M. G. (Eds.) (1976) *Transport in Plants II. A Cells B Tissues and Organs.* Springer-Verlag, Berlin, Heidelberg and New York.
MACKAY, J. F. G. and WEATHERLEY, P. E. (1973) *J. Exp. Bot.*, **24,** 15–28.
MANSFIELD, T. A. (1976) *Phil. Trans. R. Soc. Lond.*, **B 273,** 541–50.
MARTIN, E. S. and MEIDNER, H., (1972) *New Phytol.*, **71,** 1045–54.
MEES, G. C. and WEATHERLEY, P. E. (1957) *Proc. Roy. Soc.*, **B 147,** 367–80.
MEIDNER, H. (1965) *Symp. Soc. Exptl. Biol. XIX*, pp. 185–203. Cambridge U.P., Cambridge.
MEIDNER, H. and HEATH, O. V. S. (1959) *J. Exp. Bot.*, **10,** 206–19.
*MEIDNER, H. and MANSFIELD, T. A. (1968) *Physiology of Stomata.* McGraw-Hill, London.
†MEIDNER, H. and SHERRIFF, D. W. (1976) *Water and Plants*, Blackie, Glasgow and London.
MEIDNER, H. and WILMER, C. (1975) *Curr. Adv. Plant Sci.*, **17,** 1–15.
MILBURN, J. A. (1964) *The Uptake of Water and Solutes by Plant Tissues.* Ph.D. Thesis, University of Aberdeen.
MILBURN, J. A. (1970) *New Phytol.*, **69,** 133–41.
MILBURN, J. A. (1973a) *Planta (Berl.)*, **110,** 253–65.
MILBURN, J. A. (1973b) *Planta (Berl.)*, **112,** 333–42.
MILBURN, J. A. (1974) *Planta (Berl.)*, **117,** 303–19.

MILBURN, J. A. (1975a) In *Form, Structure and Function in Plants.* Sarita Prakashan, U.P., India.
*MILBURN, J. A. (1975b) In *Encyclop. Pl. Physiol.*, vol. 1, eds. Zimmermann M. H. and Milburn, J. A., Springer-Verlag, Berlin and New York.
MILBURN, J. A. and COVEY-CRUMP, P. A. K. (1971) *New Phytol.*, **70**, 427–34.
MILBURN, J. A. and DODOO, G. (in prep.).
MILBURN, J. A. and JOHNSON R. P. C. (1966) *Planta* (*Berl.*), **69**, 43–52.
MILBURN, J. A. and MCLAUGHLIN, M. E. (1974) *New Phytol.*, **73**, 861–71.
MILBURN, J. A. and WEATHERLEY, P. E. (1971) *New Phytol.*, **70**, 929–39. Springer, Berlin and New York.
MILBURN J. A. and ZIMMERMANN, M. H. (1977) *New Phytol.*, **79**, 535–41.
MITCHELL, P. (1966) *Biological Rev.*, **44**, 445–502.
MOHL. H. VON (1956) *Bot. Zeit.*, **14**, 697–704.
*MONTEITH, J. (1973) *Principles of Environmental Physics.* Arnold, London.
*MOORE, D. J. and MOLLENHAUER, H. H. (1974) In *Dynamic Aspects of Plant Ultra Structure*, pp. 84–137. Ed. Robards, A. W. McGraw-Hill, New York and London.
*MÜNCH, E. (1930) *Die Stoffbewegungen in der Pflanze.* Fischer, Jena.
MYERS, G. M. P. (1951) *J. Exp. Bot.*, **2**, 129–43.
*NEWMAN, E. I. (1976) *Pl. Physiol.*, **57**, 738–9.
*NOBEL, P. S. (1974) *Introduction to Biophysical Plant Physiology.* Freeman, San Francisco.
NOBLE-NESBITT, J. (1970) *J. Exp. Biol.*, **52**, 193–200.
NOBLE-NESBITT, J. (1975) *J. Exp. Biol.*, **62**, 657–69.
OKASHA, A. Y. K. (1971) *J. Exp. Biol.*, **55**, 435–48.
OKASHA, A. Y. K. (1972) *J. Exp. Biol.*, **57**, 285–96.
OVERBEEK, J. van (1942) *Amer. J. Bot.*, **29**, 677–83.
PASSIOURA, J. B. (1972) *Aust. J. Agric. Res.*, **23**, 745–52.
PEEL, A. (1965) *Ann. Bot.*, **29**, 119–30.
PENNY, M. G. and BOWLING D. J. F. (1974) *Planta* (*Berl.*), **119**, 17–25.
*PFEFFER, W. (1897) *The Physiology of Plants* (Transl. Ewart) **1**. Clarendon Press, Oxford.
POTTER, A. W. and MILBURN, J. A. (1970) *New Phytol.*, **69**, 961–9.
PRESTON, R. D. (1958) *Colston Papers. Proc. 10 Symp.*, **10**, 366–79. Butterworth, London.
RASCHKE, K. (1970) *Science*, **167**, 189–91.
REES, A. R. (1961) *J. Exp. Bot.*, **12**, 129–46.
ROBARDS, A. W. and ROBB, M. E. (1972) *Science*, **178**, 980–2.
ROBERTSON, J. D. (1959) *Biochem. Soc. Symp.*, **16**, 3.
RODDIE, I. A. (1971) *Physiology for Practitioners.* Churchill Livingstone, Edinburgh and London.
ROSENE, H. (1943) *Pl. Physiol.*, **18**, 588–607.
RUFZ DE LAVISON, J. (1910) *Rev. Gen. Botanique.*, **22**, 225–41.
†RUTTER, J. A. (1972) *Transpiration.* Oxford U.P., London.
SABININ, D. A. (1925) *Bull. Inst. Recherches Biologique.*, Univ. de Perm., **4** Suppl. 2, 1–136.
*SACHS, J. (1887) *Lectures on the Physiology of Plants.* (Transl. Ward, H. M.) Clarendon Press, Oxford.
SAMPSON, J. (1961) *Nature* (*Lond.*), **191**, 932.
SAUTER, J. J. (1974) *Year Book Science and Technology.* McGraw-Hill, New York.
SAUTER, J. J., ITEN, W. and ZIMMERMANN, M. H. (1973) *Can. J. Bot.*, **51**, 1–8.
SAYRE, J. D. (1946) *Ohio J. Sci.*, **26**, 233–66.
SCHOLANDER, P. F. (1968) *Physiologia Plantarum*, **21**, 251–61.
SCHOLANDER, P. F., HAMMEL, H. T., BRADSTREET, E. D. and HEMMINGSEN, E. A. (1965) *Science*, **148**, 339–46.

†SESTAK, Z., CATSKY, J. and JARVIS, P. G. (1971) *Plant Photosynthetic Production: Manual of Methods*. Junk, The Hague.
SHEIKHOLESLAM, S. N., and CURRIER, H. B. (1977) *Plant Physiol.*, **59**, 376–80.
SHERRIFF, D. W. (1972) *J. Exp. Bot.*, **23**, 1086–95.
SHERRIFF, D. W. (1973) *J. Exp. Bot.*, **24**, 641–7.
SHERRIFF, D. W. (1974) *J. Exp. Bot.*, **25**, 675–83.
*SCHERY, R. W. (1954) *Plants for Man*. Allen and Unwin, London.
*SJOSTAND, F. S., (1968) *Regulatory Functions of Biological Membranes*. Elsevier, Amsterdam.
*SLATYER, R. O. (1967) *Plant Water Relationships*, p. 258. Academic Press, London and New York.
SPANNER, D. C. (1952) *Ann. Bot. N.S.*, **16**, 133–6.
SPANNER, D. C. (1973) *J. Exp. Bot.*, **24**, 816–19.
SPRENT, J. I. (1972) *New Phytol.*, **71**, 603–11.
SQUIRE, G. R. and MANSFIELD, T. A. (1972) *New Phytol.*, **73**, 433–40.
STEUDLE, E., LUTTGE, U. and ZIMMERMANN, U. (1975) *Planta (Berl.)*, **126**, 229–46.
STEUDLE, E., ZIMMERMANN, U. and LÜTTGE, U. (1977) *Plant Physiol.*, **59**, 285–9.
STOCKER, O. (1929) *Planta (Berl.)*, **7**, 382–7.
STOCKER, O. (1933) *Jarb. wiss Bot.*, **78**, 751–856.
STOCKER, O. (1974) *Flora*, **163**, 493–529.
STOKES, R. H. (1947) *J. Amer. Chem. Soc.*, **69**, 1291.
STOCKING, C. R. and HEBER, U. (1976) *Transport in Plants III. Intracellular Interactions and Transport Processes*. Springer-Verlag, Berlin, Heidelberg and New York.
SWANSON, R. H. (1965) Seasonal course of transpiration of Lodgepole pine and Engelmann spruce, *Intern. Symp. Forest Hydrology*. Penn. State Univ. U.S.A. 417–32.
*SYDENHAM, P. H. and FINDLAY, G. P. (1975) In *Ion Transport in Plants*, ed. Anderson, W. P. Academic Press, London and New York.
TAMMES, P. M. L. (1952) *Proc. Koninkl. Ned. Akad. Wetenschap C55*, 141–3.
TANTON, T. W. and CROWDY, S. H. (1972) *J. Exp. Bot.*, **23**, 619–26.
*THAINE, J. F. (1967) *Principles of Osmotic Phenomena*. Roy. Inst. Chem. London.
THOMAS, M. D. and HILL, G. R. (1937) *Pl. Physiol.*, **12**, 285–307.
THUT, H. F. (1932) *Ohio J. Sci.*, **28**, 292–8.
TOBIESSON, P., RUNDEL, P. W. and STECKER, R. E. (1971) *Pl. Physiol.*, **48**, 303–4.
TOMLINSON, P. B. and ZIMMERMANN, M. H. (1967) *Bull. Int. Ass. Wood Anat.*, **2**, 4–24.
TURGEON, R. and WEBB, J. A., (1973) *Planta (Berl.)*, **113**, 179–91.
TYREE, M. T. and ZIMMERMANN, M. H. (1971) *J. Exp. Bot.*, **22**, 1–18.
VAN DEN HONERT, T. H. (1948) *Disc. Farad. Soc. (Lond.)*, **3**, 146.
VELKOV, D. and KALUDIN, K. (1970) *Gorskostop. Nauka*, **7**, 33–42.
*WALTER, H. (1973) *Vegetation of the Earth*. Springer, New York.
WALTER, H. and STEINER, M. (1936) *Z. Bot.*, **30**, 65–193.
†WARDLAW, I. F. and PASSIOURA, J. B. (Eds.) (1976) *Transport and Transfer Processes in Plants*. Academic Press, New York, San Francisco, London.
*WAREING, P. F. and PHILLIPS, I. D. J. (1976) *The Control of Growth and Differentiation in Plants*. Pergamon Press, Oxford.
WEATHERLEY P. E. (1950) *New Phytol.*, **48**, 81–97.
*WEATHERLEY P. E. (1963) In *The Water Relations of Plants*, pp. 85–100. Eds. Rutter and Whitehead. Blackwell, Oxford.
WEATHERLEY, P. E., PEEL, A. J. and HILL, G. P. (1959) *J. Exp. Bot.*, **10**, 1–16.
WEATHERLEY, P. E. and SLATYER, R. O. (1957) *Nature (Lond.)*, **179**, 1085–6.

WEST, D. W. and GAFF, D. F. (1976) *Planta* (*Berl.*), **129,** 15–18.
WHITE, P. R. (1938) *Amer. J. Bot.*, **25,** 223–7.
WILLIS, A. J. YEMM, E. W. and BALSUBRAMANIAN, S. (1963) *Nature* (*Lond.*), **199,** 265–6.
WRIGHT, S. T. C. (1969) *Planta* (*Berl.*), **86,** 10–20.
WRIGHT, S. T. C. and HIRON, R. W. P. (1969) *Nature* (*Lond.*), **224,** 719–20.
ZIMMERMANN, M. H. (1963) *Sci. Amer.*, **208,** 132–42.
ZIMMERMANN, M. H. (1964) *Pl. Physiol.*, **39,** 568–72.
ZIMMERMANN, M. H. (1976) In *Transport and Transfer Processes in Plants.* Ed. Wardlaw, I. F. and Passioura, J. B. Academic Press, New York.
*ZIMMERMANN, M. H. and BROWN, C. R. (1971) *Trees Structure and Function.* Springer-Verlag, Berlin.
*ZIMMERMANN, M. H. and MILBURN, J. A. (Eds.) (1975) *Transport in Plants I Phloem Transport.* Springer-Verlag, Berlin, Heidelberg and New York.
ZIMMERMANN, M. H. and TOMLINSON, P. B. (1965) *J. Arnold Arboretum.* **46,** 160–77.
ZIMMERMANN, M. H. and TOMLINSON, P. B. (1966) *Science*, **152,** 72–3.
ZIMMERMANN, M. H. and TOMLINSON, P. B. (1967) *Bull. Int. Assoc. Wood anat.* **1**.
†ZIMMERMANN, U. and DAINTY, J. (Eds.) (1974) *Membrane Transport in Plants.* Springer-Verlag, Berlin, Heidelberg and New York.
ZIMMERMANN, U. and STEUDLE, E. (1975) *Audt. J. Plant Physiol.*, **2,** 1–12.

Index